Calculus&Mathematica

INTEGRALS: Measuring Accumulated Growth

Bill Davis
Ohio State University

Horacio Porta
University of Illinois, Urbana-Champaign

Jerry Uhl
University of Illinois, Urbana-Champaign

Addison-Wesley Publishing Company

Reading, Massachusetts • Menlo Park, California • New York
Don Mills, Ontario • Wokingham, England • Amsterdam • Bonn
Sydney • Singapore • Tokyo • Madrid • San Juan • Milan • Paris

Preface

Growth and change. These are the themes of life and measuring them is the theme of calculus. Learning how to measure accumulated growth and learning how to use this measurement is what this part of Calculus&*Mathematica* is all about.

In this unit of Calculus&*Mathematica*, you'll work with the mathematics underlying measurements of accumulated growth and related measurements of area under curves. You'll see how this measurement of accumulated growth is related to the derivative. Along the way, you'll learn how to measure the acreage of an irregular plot of land, how acceleration measurements give you velocity measurements and how velocity measurements give you distance measurements. You'll learn how to measure the volume of solids by slicing and accumulating, how to measure the length of a curve, how the famous bell-shaped curve is used to break up the results of large-scale standardized tests into percentiles, and how to use the bell-shaped curve to plan bombing runs.

You'll also learn how to make sense of e^x when x is an imaginary number and you'll see why folks say

$$e^{\sqrt{-1}\,x} = \cos[x] + \sqrt{-1}\,\sin[x]$$

so that

$$e^{\sqrt{-1}\,\pi} = -1.$$

How to Use This Book

In Calculus&*Mathematica*, great care has been taken to put you in a position to learn visually. Instead of forcing you to attempt to learn by memorizing impenetrable jargon, you will be put in the position in which you will experience mathematics by seeing it happen and by making it happen. And you'll often be asked to describe what you see. When you do this, you'll be engaging in active mathematics as opposed to the passive mathematics you were probably asked to do in most other math courses. In order to take full advantage of this crucial aspect of Calculus&Mathematica, your first exposure to a new idea should be on the live computer screen where you can interact with the electronic text to your own satisfaction. This means that you should avoid "introductory lectures" and you should avoid reading this book at first. After you have some familiarity with new ideas as found on the computer screen, you should seek out others for discussion and you can refer to this book to brush up on a point or two after you leave the computer. In the final analysis, this book is nothing more than a partial record of what happens on the screen.

Once you have participated in the mathematics and science of each lesson, you can sharpen your hand skills and check up on your calculus literacy by trying the questions in the Literacy Sheet associated with each lesson. The Literacy Sheets appear at the end of each book.

Significant Changes from the Traditional Course

Student writing, plotting, and experimentation is the stock in trade of the course.

Following Courant and Artin, the integral is introduced as a measurement of area under a curve. The approach is strong enough to get to the idea of approximating by trapezoids and to get to the idea of the fundamental formula while totally avoiding the bureaucracy of Riemann sums. This results in significant saving of time for more important matters.

Riemann sums are used only occasionally and they are used only as part of the process of setting up an integral for a required measurement.

Integrals are used to measure area, length, and volume but are not used to define these notions. The issue of defining these notions is not addressed.

The indefinite integral, $\int f[x]\, dx$ (the integral without limits), is totally deemphasized and mentioned only in passing. Reason: In this course, all integrals make measurements; the indefinite integral measures nothing. On the other hand, the functions like $\int_a^x f[t]\, dt$ (as a function of x) are under heavy scrutiny.

The idea for the fundamental formula is set up experimentally by having the student take selected functions $g[x]$, putting $f[x] = \int_a^x g[t]\, dt$, and plotting

$$\frac{f[x + h] - f[x]}{h} \qquad \text{and} \qquad g[x]$$

on the same axes for small h's. This experiment not only sets up the fundamental formula but also reinforces the idea of what the derivative is.

The explanation of the fundamental formula is based on the trapezoidal rule.

Threefold emphasis on measurements made with integrals: Measurements based on slicing and accumulating, measurements based on approximations, and measurements of accumulated growth via the fundamental formula.

At the request of physics professors, the complex exponential enters an American calculus course for the first time. The matter-of-fact treatment is based in part on ideas from the Feynman Lectures.

Although traditional techniques of integration are deemphasized, integration by parts and the idea of transforming one integral into another via the formula

$$\int_a^b f'[u[x]]\, u'[x]\, dx = \int_{u[a]}^{u[b]} f'[u]\, du$$

survive. Reason: Knowledge of transformations of integrals helps to prepare the student to understand where *Mathematica*-generated formulas like

$$\int_0^x \cos[t^4]\, t\, dt = \sqrt{\pi/8}\,\text{FresnelC}[4x^2/\sqrt{8\pi}]$$

come from and what they mean.

The second reason for emphasis on transforming integrals is that simple transformations unlock the mysteries of the normal probability distribution. Most students have heard of the famous "bell-shaped curve" and want to work with it. In this course, they get their chance by using properties of the normal probability distribution to set quartiles on SAT scores and they even have the experience of "curving a test."

Double integrals as volume measurements are taken up earlier than they are in the traditional course. *Mathematica*'s ability to plot surfaces is a strong motivator for taking up this topic as soon as possible. The Gauss-Green formula (Green's theorem) is introduced for the purpose of replacing calculationally intractable double integrals, by one-dimensional integrals, which can be easily calculated numerically. Students get to try this out by using Gauss's normal law in 2D to decide, as done in the Pentagon, how many bombs to drop on a target to get a given kill probability. Students will get another chance to work with the Gauss-Green formula in vector calculus. Students appreciate the chance to have two cracks at the Gauss-Green formula.

Contents

INTEGRALS: Measuring Accumulated Growth

In normal use, the student engages in all the mathematics and the student engages in selected experiences in math and science as assigned by the individual instructor.

2.01 Integrals for Measuring Area   1

Mathematics   Integrals defined as area measurement as done in E. Artin's MAA notes written in the 1950's. Approximations by trapezoids.

Science and math experience   Integrals of functions given by data lists. Using known area formulas for triangles, trapezoids and circles to calculate integrals. Odd functions. Trying to break the code of the integral by taking selected functions $g[x]$, putting $f[x] = \int_a^x g[t]\, dt$ and plotting

$$\frac{f[x+h] - f[x]}{h} \qquad \text{and} \qquad g[x]$$

on the same axes for small h's. Plotting $f[x] = \int_0^x \cos[t]\, dt$ and guessing a formula for $f[x]$. Plotting $f[x] = \int_0^x \sin[t]\, dt$ and guessing a formula for $f[x]$. Estimating the acreage of farm field bordered by a river.

2.02 Breaking the Code of the Integral: The Fundamental Formula   41

Mathematics   If $f[t]$ is given by $f[x] = \int_a^x g[t]\, dt$, then $f'[x] = g[x]$. The fundamental formula $f[x] - f[a] = \int_a^x f'[t]\, dt$.

Science and math experience   Relating distance, velocity, and acceleration through the fundamental formula. Getting the feel of the fundamental formula by using it to calculate integrals by hand. Relating $\int_a^x g[t]\,dt$ to the solution of the differential equation $y'[x] = g[x]$ with $y[a] = 0$. Very brief look at the "indefinite integral," $\int g[x]\,dx$. Measuring area between curves. The error function, Erf[x], and other functions defined by integrals. Measurements of accumulated growth. Coloring ceramic tiles.

2.03 Measurements   97

Mathematics   Measurements based on slicing and accumulating: Area and volume; density and mass. Measurements based on approximating and measuring: Arc length. Measurements based on the fundamental formula: Accumulated growth.

Science and math experience   Volumes of solids with no special emphasis on solids of rotation. Volume measurements of curved tubes and horns. Eyeball and precise estimates of curve lengths. Filling water tanks. Harvesting corn. Voltage drop. Another look at linear dimension. Work. Present value of a profit-making scheme. Catfish harvesting. Designing an 8-fluid-ounce logarithmic champagne glass.

2.04 Transforming Integrals   141

Mathematics   Using the chain rule and the fundamental formula to see why

$$\int_a^b f'[u[x]]\ u'[x]\,dx = \int_{u[a]}^{u[b]} f'[u]\,du$$

and using this fact to transform one integral into another. Measuring area under curves given parametrically. Bell shaped curves and Gauss's normal probability law; mean and standard deviation.

Science and math experience   Study of the error function, Erf[x]. Using transformations to explain *Mathematica* output. Polar plots and area measurements. Using transformations to explain the meaning of standard deviation in Gauss's normal law. Expected life of light bulbs and how long to set the guarantee on them. Using Gauss's normal law to help to program coin-operated coffee machines. IQ test results. Using Gauss's normal law to organize SAT scores into quartiles and deciles. Comparison of 1967 and 1987 SAT scores. "Grading on the curve."

2.05 2D Integrals and the Gauss-Green Formula   195

Mathematics Meaning of the plot of $z = f[x, y]$. The 2D integral $\iint_R f[x, y]\, dx\, dy$ as a volume measurement via slicing and accumulating. Gauss-Green formula (Green's theorem) as a way of calculating a double integral numerically as a single integral.

Science and math experience Volume and area measurements with 2D integrals. Area and volume measurements via the Gauss-Green formula. Average value and centroids. Calculation strategies. Plotting and measuring. Gauss's normal law in 2D and using it, as done in the Pentagon, to decide how many bombs to drop on a target.

2.06 More Tools and Measurements   237

Mathematics Separating the variables and integrating to get formulas for the solutions of some differential equations. Integration by parts. Complex numbers and the complex exponential $e^{s+i_t} = e^s(\cos[t] + i\sin[t])$.

Science and math experience Formulas for the solutions of the differential equations involved in the chemical model and the spread of infection model. Hyperbolic functions and their relation to trigonometric functions. Using the complex exponential to help to understand the *Mathematica* output from the Solve instruction. Gamma function. Integration by parts and integration by iteration. Error propagation in forward iteration. Error reduction by backwards iteration.

2.07 Traditional Pat Integration Procedures for Special Situations   277

This lesson exists only in electronic form for those with a special interest in how to begin to program a computer to do integrals.

Mathematics Undetermined coefficients. Complex numbers and partial fractions. Wild card substitutions with the help of a trigonometric, hyperbolic, or ad hoc function. Integration by parts.

Science and math experience Not much, although the experience gained from trying the method of undetermined coefficients is good experience in setting up and solving systems of linear equations.

LESSON 2.01

Integrals for Measuring Area

Basics

■ B.1) $\int_a^b f[x]\,dx$ measures area

Calculus&Mathematica is pleased to acknowledge the heavy influence of Emil Artin's book *Calculus with Analytic Geometry* (notes by G. B. Seligman), Committee on the Undergraduate Program, Mathematical Association of America, 1957, on this problem. Although dated in spots, this short book remains a jewel.

Here is the idea of the integral: Take a function $y = f[x]$ and plot it on an interval $[a, b]$. It might look something like this:

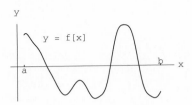

Look at the area between the graph of $y = f[x]$ and the x-axis:

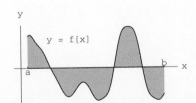

If a part of this area lies above the x-axis, call it positive. If a part lies below the x-axis, call it negative. This gives the signs as shown below:

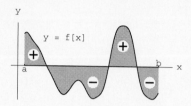

Now define a number that all folks call the integral of $f[x]$ from a to b. You signify this number by writing $\int_a^b f[x]\,dx$ and you calculate its value by summing up the measurements of these areas taken with the corresponding signs. Accordingly, if $x1 < x2 < x3$ are the points between a and b at which the curve crosses the x-axis as shown below:

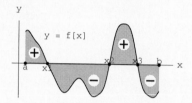

Then $\int_a^b f[x]\,dx$ is the sum of four measurements:

$$\int_a^{x1} f[x]\,dx + \int_{x1}^{x2} f[x]\,dx + \int_{x2}^{x3} f[x]\,dx + \int_{x3}^{b} f[x]\,dx$$

where $\int_a^{x1} f[x]\,dx$ and $\int_{x2}^{x3} f[x]\,dx$ are positive and $\int_{x1}^{x2} f[x]\,dx$ and $\int_{x3}^b f[x]\,dx$ are negative.

B.1.a) Given a specific function $f[x]$ and given specific numbers a and b with $a \leq b$, when you calculate $\int_a^b f[x]\,dx$, do you get a number or do you get another function?

Answer: As a signed measurement of area, $\int_a^b f[x]\,dx$ is a number.

B.1.b.i) What does

$$\int_1^4 \left(\frac{x}{2} + 1\right)\,dx$$

measure?

Answer: $\int_1^4 (x/2+1)\, dx$ measures the area under the plot of $f[x] = (x/2+1)$ and above the segment $[1, 4]$ on the x-axis. Here's the plot:

In[1]:=
```
Clear[f,x]
f[x_] = (x/2 + 1);
Plot[f[x],{x,1,4},
PlotRange->{{0,5},{0,4}},
AspectRatio->Automatic,
AxesLabel->{"x","y"}];
```

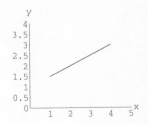

Here's a picture of the area measured by $\int_1^4 (x/2+1)\, dx$:

In[2]:=
```
FilledPlot[f[x],{x,1,4},
AxesLabel->{"x","y"}];
```

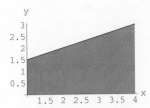

This integral is the area of a trapezoid with:

In[3]:=
```
shortHeight = f[1]; tallHeight  = f[4]; width = 3;
```

Consequently, the area measurement $\int_1^4 f[x]\, dx = \int_1^4 (x/2+1)\, dx$ in square units is given by:

In[4]:=
```
width ((tallHeight + shortHeight)/2)
```

Out[4]=

$\dfrac{27}{4}$

Mathematica can also handle this measurement directly:

In[5]:=
```
Integrate[(x/2 + 1),{x,1,4}]
```

Out[5]=

$\dfrac{27}{4}$

Good going.

B.1.b.ii)

What does

$$\int_{-4}^{4} \left(\frac{x}{2} + 1\right) dx$$

measure?

Answer: Here is a plot:

In[6]:=
```
Clear[f,x]
f[x_] = (x/2 + 1);
Plot[f[x],{x,-4,4},
AspectRatio->Automatic,
AxesLabel->{"x","y"}];
```

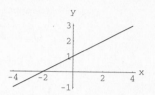

Here's a picture of the area measured by $\int_{-4}^{4} (x/2 + 1) \, dx$:

In[7]:=
```
FilledPlot[f[x],{x,-4,4},
AxesLabel->{"x","y"}];
```

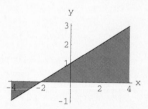

Better find out where the area goes from positive to negative:

In[8]:=
```
Solve[f[x] == 0]
```

Out[8]=
```
{{x -> -2}}
```

The measurement $\int_{-4}^{-2} (x/2 + 1) \, dx$ of the triangular area on the left is negative because it is below the x-axis:

In[9]:=
```
height = f[-4]; base = 2;
triangleleft = (1/2) base height
```

Out[9]=
```
-1
```

Now look at the triangle on the right:

In[10]:=
```
FilledPlot[f[x],{x,-4,4},
AxesLabel->{"x","y"}];
```

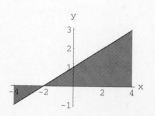

The measurement $\int_{-2}^{4}(x/2+1)\,dx$ of the triangular area on the right is positive:

In[11]:=
```
height = f[4]; base = 6;
triangleright = (1/2) base height
```
Out[11]=
```
9
```

Consequently,

$$\int_{-4}^{4}\left(\frac{x}{2}+1\right)\,dx=\int_{-4}^{-2}\left(\frac{x}{2}+1\right)\,dx+\int_{-2}^{4}\left(\frac{x}{2}+1\right)\,dx$$

is given by:

In[12]:=
```
triangleleft + triangleright
```
Out[12]=
```
8
```

Mathematica can calculate $\int_{-4}^{4}(x/2+1)\,dx$ directly:

In[13]:=
```
Integrate[(x/2 + 1),{x,-4,4}]
```
Out[13]=
```
8
```

Got it, and it wasn't very hard.

■ B.2) Three properties of integrals

$$\rightarrow\quad\int_{a}^{b}f[x]\,dx=\int_{a}^{c}f[x]\,dx+\int_{c}^{b}f[x]\,dx\quad\text{for any number }c\text{ with }a<c<b;$$

$$\rightarrow\quad\int_{a}^{b}Kf[x]\,dx=K\int_{a}^{b}f[x]\,dx\quad\text{for any number }K;$$

$$\rightarrow\quad\int_{a}^{b}f[x]\,dx=\int_{a}^{b}f[t]\,dt.$$

Calculus&*Mathematica* is again pleased to acknowledge the heavy influence of Emil Artin's book *Calculus with Analytic Geometry* (notes by G. B. Seligman), Committee on the Undergraduate Program, Mathematical Association of America, 1957, on this problem.

B.2.a)

Explain the formula

$$\int_{a}^{b}f[x]\,dx=\int_{a}^{c}f[x]\,dx+\int_{c}^{b}f[x]\,dx$$

for any number c with $a<c<b$.

Answer: The integral $\int_a^b f[x]\,dx$ measures the (signed) area between the graph of $y = f[x]$ and the segment of the x-axis between $x = a$ and $x = b$.

The sum $\int_a^c f[x]\,dx + \int_c^b f[x]\,dx$ measures the same signed area but calculates it in two parts. Try it for $f[x] = 2x - 4$, $a = 0$, $c = 3$, and $b = 5$. The number $\int_a^b f[x]\,dx$ is given by:

In[14]:=
```
Clear[f,x]; f[x_] = 2 x - 4; a = 0; c = 3; b = 5;
everything = Integrate[f[x],{x,a,b}]
```
Out[14]=
```
5
```

This should be the same as $\int_a^c f[x]\,dx + \int_c^b f[x]\,dx$:

In[15]:=
```
leftpart = Integrate[f[x],{x,a,c}]; rightpart = Integrate[f[x],{x,c,b}];
leftpart + rightpart
```
Out[15]=
```
5
```

Right on the money. Try it for another c between a and b:

In[16]:=
```
c = 1;
leftpart = Integrate[f[x],{x,a,c}]; rightpart = Integrate[f[x],{x,c,b}];
leftpart + rightpart
```
Out[16]=
```
5
```

On the money again. And it will stay on the money for any function $f[x]$ and any a, b, and c as long as they line up with $a < c < b$.

B.2.b) | Explain the formula
$$\int_a^b K\,f[x]\,dx = K \int_a^b f[x]\,dx$$
for any number K.

Answer: The expression $K \int_a^b f[x]\,dx$ measures an area and then multiplies by K to expand the value of the area. The expression $\int_a^b K\,f[x]\,dx$ expands the area vertically first (by multiplying by K) and then measures the result. Either expression leads to the same result. Try it for $f[x] = 4x + 1$, $a = 1$, and $b = 7$:

In[17]:=
```
Clear[K,f,x]; f[x_] = 4 x + 1; a = 1; b = 7;
Integrate[K f[x],{x,a,b}]
```
Out[17]=
```
102 K
```

This should agree with:

In[18]:=
```
K Integrate[f[x],{x,a,b}]
```

Out[18]=
```
102 K
```

It does. Try it for a particular value of K, another function, and other choices of a and b with $a < b$:

In[19]:=
```
Clear[K,f,x]; f[x_] = 8 - 2 x;
a = 3; b = 12; K = Pi;
Integrate[K f[x],{x,a,b}]
```

Out[19]=
```
-63 Pi
```

This should agree with:

In[20]:=
```
K Integrate[f[x],{x,a,b}]
```

Out[20]=
```
-63 Pi
```

This will happen for any function $f[x]$, any constant K, and any a and b as long as they line up with $a < b$.

B.2.c) Explain the formula

$$\int_a^b f[x]\,dx = \int_a^b f[t]\,dt.$$

Answer: Set up an example:

In[21]:=
```
Clear[f,x,t]
f[x_] = E^(-x/2) Cos[3 x];
a = 0.8; b = 3.7;
Clear[s]
FilledPlot[f[s],{s,a,b},
AxesLabel->{"s","f[s]"}];
```

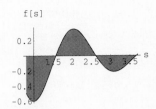

$\int_a^b f[x]\,dx$, $\int_a^b f[t]\,dt$, and $\int_a^b f[s]\,ds$ measure the same signed area, and no matter how you label the horizontal axis, the measurement is the same:

In[22]:=
```
Integrate[f[s],{s,a,b}]
```

Out[22]=
```
-0.225169
```

In[23]:=
```
Integrate[f[x],{x,a,b}]
```
Out[23]=
```
-0.225169
```
In[24]:=
```
Integrate[f[t],{t,a,b}]
```
Out[24]=
```
-0.225169
```

This will work for all functions. The use of the symbols s or t instead of the symbol x is a matter of bureaucracy and not a matter of mathematics or science. Only bean counters want to worry about it.

■ B.3) Integration by approximation by trapezoids and the *Mathematica* instruction NIntegrate

B.3.a.i)   Plot on the same axes:

$$f[x] = \sin[\pi x] \qquad \text{for } 0 \le x \le 1$$

and the broken line that joins the five points on the curve

$$\{0, f[0]\}, \ \left\{\frac{1}{4}, f\left[\frac{1}{4}\right]\right\}, \ \left\{\frac{1}{2}, f\left[\frac{1}{2}\right]\right\}, \ \left\{\frac{3}{4}, f\left[\frac{3}{4}\right]\right\}, \ \{1, f[1]\}.$$

> Measure the area under the broken line segments over $0 \le x \le 1$ and discuss why this measurement gives a rough estimate of the area measurement $\int_0^1 f[x]\,dx$.

Answer: Define the function:

In[25]:=
```
Clear[f,x]; f[x_] = Sin[Pi x];
```

Look at the plot:

In[26]:=
```
a = 0; b = 1;
curveplot = Plot[f[x],{x,a,b},
PlotStyle->{{Red,Thickness[0.01]}},
AxesLabel->{"x","f[x]"},
PlotRange->{0,1.2},
AxesOrigin->{0,0}];
```

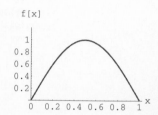

$\int_a^b f[x]\,dx$ measures the area under the curve and over the interval $a \le x \le b$ on the x-axis. In this plot, $a = 0$ and $b = 1$. Now make a table of the points

$$\{0, f[0]\}, \ \left\{\frac{1}{4}, f\left[\frac{1}{4}\right]\right\}, \ \left\{\frac{1}{2}, f\left[\frac{1}{2}\right]\right\}, \ \left\{\frac{3}{4}, f\left[\frac{3}{4}\right]\right\}, \ \{1, f[1]\}.$$

The jump in the x-coordinate from point to point is:

In[27]:=
```
jump = (b - a)/4
```

Out[27]=

$$\frac{1}{4}$$

So the points are given by:

In[28]:=
```
points = Table[{x,f[x]},{x,a,b,jump}]
```

Out[28]=

$$\{\{0, 0\}, \ \{\frac{1}{4}, \ \frac{1}{\text{Sqrt}[2]}\}, \ \{\frac{1}{2}, \ 1\}, \ \{\frac{3}{4}, \ \frac{1}{\text{Sqrt}[2]}\}, \ \{1, \ 0\}\}$$

Set up the plot of the points and the broken line segments:

In[29]:=
```
pointplot = ListPlot[points,PlotStyle->
{Red,PointSize[0.03]},DisplayFunction->Identity];
brokenlineplot =
Graphics[{Thickness[0.01],Line[points]}];
Show[curveplot,pointplot,brokenlineplot,
AxesLabel->{"x",""},PlotRange->{0,1.2},
AxesOrigin->{0,0}];
```

The area underneath the broken line segments takes a very healthy bite of the area under the curve. The (signed) area between the x-axis and the broken line segments plotted above is the sum of the areas of consecutive trapezoids based on segments $[x, x + \text{jump}]$ on the x-axis:

In[30]:=
```
Clear[trapezoid,x]
trapezoid[x_] := Graphics[{Thickness[0.01],
Line[{{x,0},{x + jump,0},{x + jump,f[x + jump]},
{x,f[x]},{x,0}}]}];
trapezoids = Table[trapezoid[x],{x,a,b - jump,jump}];
Show[curveplot,pointplot,trapezoids,
AxesLabel->{"x",""},PlotRange->{0,1.2},
AxesOrigin->{0,0}];
```

Each trapezoid has a base whose length equals jump. Each has two heights—namely, $f[x]$ and $f[x + \text{jump}]$. So the area of each of the trapezoids is

$$\text{jump} \ \frac{f(x) + f(x + \text{jump})}{2}$$

and you start at $x = 0 = a$ and finish at $x = 1 - \text{jump} = b - \text{jump}$. The sum of the area of these four trapezoids is:

In[31]:=
```
Area[4] = N[Sum[jump (f[x] + f[x + jump])/2, {x,a,b - jump,jump}]]
```
Out[31]=
```
0.603553
```

This sum measures the sum of areas of the following four trapezoids:

In[32]:=
```
Table[Show[curveplot,pointplot,trapezoid[x],
AxesLabel->{"x",""}],{x,a,b - jump,jump}];
```

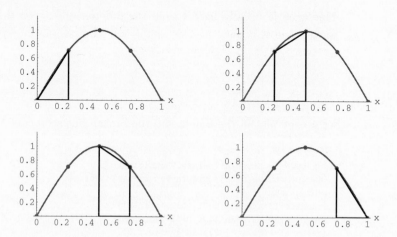

The total area under the broken line segments over $0 \le x \le 1$ is the sum of the areas of the trapezoids:

In[33]:=
```
Area[4] = N[Sum[jump (f[x] + f[x + jump])/2, {x,a,b - jump,jump}]]
```
Out[33]=
```
0.603553
```

This is a fairly reasonable rough estimate of the true value of the area measurement $\int_a^b f[x]\,dx$. And it wasn't all that hard to get!

B.3.a.ii) | How do you get a better estimate than the estimate in part B.3.a.i)?

Answer: Use a smaller jump; the old jump was $(b - a)/4 = 1/4$. Go with

$$\text{jump} = \frac{b - a}{10} = \frac{1}{10} :$$

In[34]:=
```
jump = (b - a)/10;
points = Table[{x,f[x]},{x,a,b,jump}];
pointplot = ListPlot[points,
PlotStyle->{Red,PointSize[0.03]},
DisplayFunction->Identity]; brokenlineplot =
Graphics[{Thickness[0.01],Line[points]}];
Show[curveplot,pointplot,brokenlineplot,
AxesLabel->{"x",""},PlotRange->{0,1.2},
AxesOrigin->{0,0}];
```

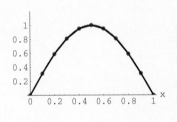

This time it's really hard to tell the difference between the area underneath the broken line segments over the x-axis and the true area $\int_a^b f[x]\,dx$ under the curve and over the x-axis. Here are the corresponding trapezoids:

In[35]:=
```
Clear[trapezoid,x]
trapezoid[x_] := Graphics[{Thickness[0.01],
Line[{{x,0},{x + jump,0},
{x + jump,f[x + jump]},{x,f[x]},{x,0}}}]];
trapezoids = Table[trapezoid[x],{x,a,b - jump,jump}];
Show[curveplot,pointplot,trapezoids,
AxesLabel->{"x",""},PlotRange->{0,1.2},
AxesOrigin->{0,0}];
```

The sum of the area inside these 10 trapezoids is:

In[36]:=
```
Area[10] = N[Sum[jump (f[x] + f[x + jump])/2, {x,a,b - jump,jump}]]
```

Out[36]=
```
0.631375
```

Compare the old estimate with the new estimate:

In[37]:=
```
{Area[4],Area[10]}
```

Out[37]=
```
{0.603553, 0.631375}
```

Of the two, the second estimate is the more trustworthy, and the plots tell you why. To go for even more accuracy, decrease the jump size again:

In[38]:=
```
jump = (b - a)/50;
points = Table[{x,f[x]},{x,a,b,jump}];
pointplot = ListPlot[points,PlotStyle->
{Red,PointSize[0.03]},DisplayFunction->Identity];
brokenlineplot = Graphics[{Thickness[0.01],
Line[points]}];Show[curveplot,pointplot,
brokenlineplot,AxesLabel->{"x",""},
PlotRange->{0,1.2},AxesOrigin->{0,0}];
```

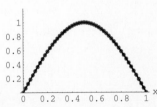

This time the points are packed so densely that it's really hard to tell the difference between the area underneath the broken line segments over the x-axis and the true area $\int_a^b f[x]\,dx$ under the curve over the x-axis. Here are the trapezoids:

In[39]:=
```
trapezoids = Table[trapezoid[x],
{x,a,b - jump,jump}];
Show[curveplot,pointplot,trapezoids,
AxesLabel->{"x",""},PlotRange->{0,1.2},
AxesOrigin->{0,0}];
```

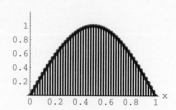

The sum of the area measurements of these 50 trapezoids is:

In[40]:=
```
Area[50] = N[Sum[jump (f[x] + f[x + jump])/2, {x,a,b - jump,jump}]]
```

Out[40]=
```
0.63641
```

Compare the old estimates with the new estimate:

In[41]:=
```
{Area[4],Area[10],Area[50]}
```

Out[41]=
```
{0.603553, 0.631375, 0.63641}
```

Of the three, the last estimate is the most trustworthy, and the plot tells you why. You can say with some confidence that $\int_a^b f[x]\,dx = 0.6$ to one accurate decimal. If you want a better estimate, look at:

In[42]:=
```
jump = (b - a)/200;
Area[200] = N[Sum[jump (f[x] + f[x + jump])/2, {x,a,b - jump,jump}]]
```

Out[42]=
```
0.636607
```

This is even more trustworthy than the estimate immediately above. You can say with even more confidence that $\int_a^b f[x]\,dx = 0.64$ to two accurate decimals. See what *Mathematica* thinks:

In[43]:=
```
N[Integrate[f[x],{x,a,b}]]
```

Out[43]=
```
0.63662
```

That last estimate was really close!

B.3.a.iii)   How are these calculations related to the *Mathematica* instruction NIntegrate?

Answer: The *Mathematica* instruction NIntegrate uses a professionally written approximation procedure similar in spirit to what you just did with the trapezoids. Let's try this instruction to test our estimates above:

In[44]:=
```
MathematicaEstimate = NIntegrate[f[x],{x,a,b}]
```

Out[44]=
```
0.63662
```

Compare this to the estimates you got above:

In[45]:=
```
{Area[4],Area[10],Area[50],Area[200],MathematicaEstimate}
```

Out[45]=
```
{0.603553, 0.631375, 0.63641, 0.636607, 0.63662}
```

The estimate coming from 200 trapezoids was pretty darned good. The advantages of the *Mathematica* instruction are:

→ Little typing.

→ Lots of accuracy.

B.3.b)   Use trapezoids to get a reasonable idea of the value of the area measurement

$$\int_1^4 \left(x^3 - 2\,x^2\right) e^{-x}\,dx.$$

Check your estimate with the *Mathematica* instruction NIntegrate.

Answer: Define the function:

In[46]:=
```
a = 1; b = 4; Clear[f,x]
f[x_] = (x^3 - 2 x^2) E^(-x)
```

Out[46]=
```
         2    3
  -2 x  + x
  -----------
       x
      E
```

Look at the plot:

In[47]:=
```
curveplot = Plot[f[x],{x,a,b},
PlotStyle->{{Red,Thickness[0.01]}},
AxesLabel->{"x",""},PlotRange->All,
AxesOrigin->{a,0}];
```

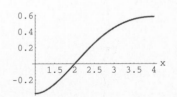

Here are the trapezoids for jump $= (b - a)/4$:

In[48]:=
```
jump = (b - a)/4;
points = Table[{x,f[x]},{x,a,b,jump}];
pointplot = ListPlot[points,PlotStyle->
{Red,PointSize[0.03]},DisplayFunction->Identity];
Clear[trapezoid,x]; trapezoid[x_] := Graphics[
{Thickness[0.01],Line[{{x,0},{x + jump,0},
{x + jump,f[x + jump]},{x,f[x]},{x,0}}]}];
trapezoids = Table[trapezoid[x],{x,a,b - jump,jump}];
Show[curveplot,trapezoids,pointplot,
AxesLabel->{"x",""},PlotRange->All,AxesOrigin->{a,0}];
```

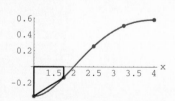

$\int_a^b f[x]\,dx$ is roughly the sum of the (signed) measurements of areas within the consecutive trapezoids. Each trapezoid is based on interval $[x, x + \text{jump}]$ on the x-axis. So each trapezoid has base $=$ jump, and each trapezoid has two heights—namely, $f[x]$ and $f[x + \text{jump}]$. So the area of each of the trapezoids is

$$\text{jump}\,\frac{f(x) + f(x + \text{jump})}{2}.$$

The sum of these areas is:

In[49]:=
```
Area[4] = N[Sum[jump (f[x] + f[x + jump])/2, {x,a,b - jump,jump}]]
```

Out[49]=
```
0.558391
```

Note how the leftmost trapezoid carries a negative area:

In[50]:=
```
N[jump (f[a] + f[a + jump])/2]
```

Out[50]=
```
-0.187847
```

This is the way it should be, because that trapezoid runs below the x-axis:

In[51]:=
```
Show[curveplot,pointplot,trapezoid[a],
AxesLabel->{"x",""},PlotRange->All,
AxesOrigin->{a,0}];
```

To get more trustworthy estimates, decrease the jump:

In[52]:=
```
jump = (b - a)/10;
Area[10] = N[Sum[jump (f[x] + f[x + jump])/2, {x,a,b - jump,jump}]]
```

Out[52]=
```
0.558858
```

Again and again:

In[53]:=
```
jump = (b - a)/100;
Area[100] = N[Sum[jump (f[x] + f[x + jump])/2, {x,a,b - jump,jump}]]
```

Out[53]=
```
0.558869
```

In[54]:=
```
jump = (b - a)/200;
Area[200] = N[Sum[jump (f[x] + f[x + jump])/2, {x,a,b - jump,jump}]]
```

Out[54]=
```
0.558869
```

Compare the estimates:

In[55]:=
```
{Area[4],Area[10],Area[100],Area[200]}
```

Out[55]=
```
{0.558391, 0.558858, 0.558869, 0.558869}
```

This is pretty convincing. A pitcher of golden liquid refreshment is bet on the claim that $\int_a^b f[x]\,dx = 0.558869$ to six accurate decimals. Check it out:

In[56]:=
```
MathematicaEstimate = NIntegrate[f[x],{x,a,b}]
```

Out[56]=
```
0.558869
```

That was a safe bet. Get out of here and collect your winnings.

Tutorials

■ T.1) Calculation of integrals for area measurements

T.1.a)

Make the area measurement

$$\int_{-1}^{5} (5.89 - 1.9\,x)\,dx.$$

Answer: To see what's happening, plot:

In[1]:=
```
Clear[f,x]
f[x_] = 5.89 - 1.9 x;
Plot[f[x],{x,-1,5},
AxesLabel->{"x","f[x]"}];
```

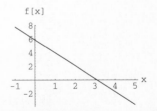

Here is a picture of the signed area measured by $\int_{-1}^{5} f[x]\, dx$:

In[2]:=
```
FilledPlot[f[x],{x,-1,5},
 AxesLabel->{"x","f[x]"}];
```

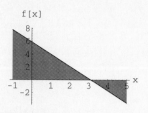

Big positive area on the left and a small negative area on the right.

Better find out where the area goes from positive to negative:

In[3]:=
```
Solve[f[x] == 0]
```

Out[3]=
```
{{x -> 3.1}}
```

The measurement $\int_{-1}^{3.1} f[x]\, dx$ of the triangular area on the left is positive because it is above the x-axis:

In[4]:=
```
height = f[-1]; base = 3.1 - (-1);
triangleleft = (1/2) base height
```

Out[4]=
```
15.9695
```

The measure $\int_{3.1}^{5} f[x]\, dx$ of the triangular area on the right is negative because it is below the x-axis:

In[5]:=
```
height = f[5]; base = 5 - 3.1;
triangleright = (1/2) base height
```

Out[5]=
```
-3.4295
```

Consequently, the area measurement

$$\int_{-1}^{5} f[x]\, dx = \int_{-1}^{3.1} f[x]\, dx + \int_{3.1}^{5} f[x]\, dx$$

is given by:

In[6]:=
```
triangleleft + triangleright
```

Out[6]=
```
12.54
```

Mathematica can make the area measurement $\int_{-1}^{5} f[x]\, dx$ directly:

In[7]:=
```
NIntegrate[f[x],{x,-1,5}]
```

Out[7]=
12.54

Piece of cake.

T.1.b)

Make the area measurement

$$\int_{-3}^{3} \sqrt{9 - x^2}\ dx.$$

Answer: This integral measures the area of the top half of the circle $x^2 + y^2 = 9$. If you don't believe this, then just solve $x^2 + y^2 = 9$ for y:

In[8]:=
```
Clear[x,y]
ysolved = Solve[x^2 + y^2 == 9,y]
```

Out[8]=

```
{{y -> Sqrt[9 - x ]}, {y -> -Sqrt[9 - x ]}}
             2                      2
```

The function you are integrating is $\sqrt{9 - x^2}$. Here's a true scale plot:

When you are plotting circles, you should always plot them in true scale so they don't look like ellipses. The option AspectRatio->Automatic guarantees a true scale plot.

In[9]:=
```
FilledPlot[Sqrt[9 - x^2], {x,-3,3},
AspectRatio->Automatic];
```

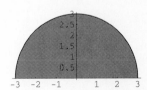

Now you know that $\int_{-3}^{3} \sqrt{9 - x^2}\ dx$ measures half the area of a circle of radius 3. So $\int_{-3}^{3} \sqrt{9 - x^2}\ dx$ is given by:

In[10]:=
```
(1/2) Pi 3^2
```

Out[10]=
$$\frac{9\ Pi}{2}$$

See what *Mathematica* does with this integral:

In[11]:=
```
Integrate[Sqrt[9 - x^2],{x,-3,3}]
```

Out[11]=
$$\frac{9\ Pi}{2}$$

Got it.

T.1.c)   Make the area measurement

$$\int_{-\pi}^{\pi} x^2 \sin[3\,x]\,dx.$$

Answer: To see what's happening, plot:

In[12]:=
```
Clear[f,x]
f[x_] = x^2 Sin[3 x];
Plot[f[x],{x,-Pi,Pi},
   AxesLabel->{"x","f[x]"}];
```

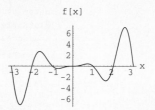

Here is a picture of the signed area measured by $\int_{-\pi}^{\pi} x^2 \sin[3\,x]\,dx$:

In[13]:=
```
FilledPlot[f[x],{x,-Pi,Pi},
   AxesLabel->{"x","f[x]"}];
```

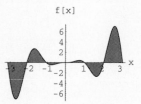

The positive and negative areas cancel each other! This happens because:

In[14]:=
```
f[-x] == -f[x]
```

Out[14]=
```
True
```

Because the positive and negative areas cancel each other, you can say with complete confidence that

$$\int_{-\pi}^{\pi} x^2 \sin[3\,x]\,dx = 0.$$

Check:

In[15]:=
```
Integrate[x^2 Sin[3 x],{x,-Pi,Pi}]
```

Out[15]=
```
0
```

There was never a doubt.

■ T.2) Areas suggested by data lists

Here are some rather simple data points and a plot of them:

```
In[16]:=
  datapoints = {{0,19.0},{7,16.1},
  {10,40.8},{13,48.2},{16,40.5},
  {19,35.9},{25,43.6}};
  dataplot = ListPlot[datapoints,
  PlotStyle->{Red,PointSize[0.04]},
  AxesLabel->{"x",""},PlotRange->{0,50},
  AxesOrigin->{0,0}];
```

T.2.a) In courses other than mathematics, you might hear professors telling you to estimate the area above the x-axis and under the curve $f[x]$ that runs through given data points such as the above. In this case, this amounts to calculating $\int_0^{25} f[x]\,dx$.

> What's the trouble with this?

Answer: There's big trouble because there are many, many functions whose plots hit all these points. One favorite of the people who talk about math but don't do any math is the unique sixth degree polynomial passing through these seven points and its plot:

```
In[17]:=
  Clear[f,x]
  f[x_] = N[InterpolatingPolynomial[datapoints,x]];
  fplot = Plot[f[x],{x,0,25},
  PlotStyle->Thickness[0.01],
  DisplayFunction->Identity];
  Show[fplot,dataplot,AxesLabel->{"x",""},
  PlotRange->All,AxesOrigin->{0,0},
  DisplayFunction->$DisplayFunction];
```

There's nothing in the data to indicate that the function should go negative; so this function lacks credibility. To make matters even worse, you could add to this function any function that is 0 at each data point. There are many functions whose plots go through the data.

T.2.b) > If you find yourself in the situation described above, what options do you have?

Answer: Your first option is to tell the professor that he doesn't know what he's talking about. This is not a wise option. Your second option is to try to come up with a reasonable function that does the job. Here's one that *Mathematica* knows about.

```
In[18]:=
  Clear[f,x]
  f[x_] = Interpolation[datapoints][x];
  fplot = Plot[f[x],{x,0,25},
  PlotStyle->Thickness[0.015],
  DisplayFunction->Identity];
  Show[fplot,dataplot,AxesLabel->{"x",""},
  PlotRange->All,AxesOrigin->{0,0},
  DisplayFunction->$DisplayFunction];
```

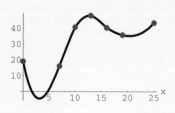

This is not so good because of the negative dip on the left. One natural guess is the function that runs on straight line segments connecting the points. Here's *Mathematica* doing that.

```
In[19]:=
  Clear[f,x]; f[x_] = Interpolation[datapoints,
  InterpolationOrder->1][x];
  fplot = Plot[f[x],{x,0,25},
  PlotStyle->Thickness[0.015],
  DisplayFunction->Identity];
  Show[fplot,dataplot,AxesLabel->{"x",""},
  PlotRange->{0,50},AxesOrigin->{0,0},
  DisplayFunction->$DisplayFunction];
```

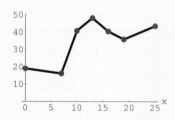

This is better than anything seen so far. It is what you get when you string line segments between the consecutive points. Go after $\int_0^{25} f[x]\,dx$:

```
In[20]:=
  NIntegrate[f[x],{x,0,25}]
NIntegrate::slwcon:
  Numerical integration converging too slowly; suspect one
    of the following: singularity, oscillatory integrand,
    or insufficient WorkingPrecision.
NIntegrate::ncvb:
  NIntegrate failed to converge to prescribed accuracy
    after 7 recursive bisections in x near x = 10.0586.
```

Out[20]=
827.85

Try again:

```
In[21]:=
  NIntegrate[f[x],{x,0,25},AccuracyGoal->1]
NIntegrate::slwcon:
  Numerical integration converging too slowly; suspect one
    of the following: singularity, oscillatory integrand,
    or insufficient WorkingPrecision.
```

Out[21]=
827.824

This means *Mathematica* advertises $\int_0^{25} f[x]\,dx = 827.8$ to one accurate decimal. Because you don't have a lot of data, it is hard to argue that more accuracy is in order; so you report $\int_0^{25} f[x]\,dx = 827.8$ and go on to something else.

T.2.c) Here are some data points and a plot:

In[22]:=
```
datapoints = {{0.00, 0.21},{0.22, 0.26},
{0.51, 0.33}, {0.89, 0.42},{1.26, 0.55},
{1.56, 0.75}, {1.92, 1.04},{2.23, 1.45},
{2.53, 1.94}, {2.91, 2.28},{3.10, 2.70},
{3.43, 3.97}, {3.79, 4.88},{4.11, 5.52},
{4.43, 6.00}, {4.76, 6.32},{5.09, 6.56},
{5.43, 6.74},{5.79, 6.88},{6.00, 6.92}};
dataplot = ListPlot[datapoints,PlotStyle->
{RGBColor[1,0,0],PointSize[.03]},AxesLabel->
{"x",""},PlotRange->{-0.5,7.5},AxesOrigin->{0,0}];
```

These data indicate measurements $\{x, y\}$ on a process that outputs y's as a function of inputs x. Apparently y increases as x increases. This time you've got closely packed data; so you expect to be able to get a fairly accurate estimate of $\int_0^6 f[x]\,dx$ for a reasonable function $f[x]$ running through the data points.

> Do it.

Answer: You can go for the function $f[x]$ whose plot runs on straight line segments connecting the given data points:

In[23]:=
```
Clear[f,x]
f[x_] = Interpolation[datapoints,
InterpolationOrder->1][x];
fplot = Plot[f[x],{x,0,6},
PlotStyle->Thickness[0.015],
DisplayFunction->Identity];
Show[fplot,dataplot,AxesLabel->{"x",""},
PlotRange->All,AxesOrigin->{0,0},
DisplayFunction->$DisplayFunction];
```

You can try:

In[24]:=
```
NIntegrate[f[x],{x,0,6}]
```
```
NIntegrate::ncvb:
   NIntegrate failed to converge to prescribed accuracy
      after 7 recursive bisections in x near x = 4.42969.
```

Out[24]=
```
19.6661
```

Try again:

In[25]:=
```
firstestimate = NIntegrate[f[x],{x,0,6},AccuracyGoal->1]
```

Out[25]=
```
19.6661
```

This tells you that $\int_0^6 f[x]\,dx = 19.7$ within 0.1. This is certainly more accuracy than the data points justify. It would be safe to say that $\int_0^6 f[x]\,dx$ is near 19.67. Another good choice is to go with the default *Mathematica* function whose plot connects the data points:

In[26]:=
```
Clear[f,x]
f[x_] = Interpolation[datapoints][x];
fplot = Plot[f[x],{x,0,6},
PlotStyle->Thickness[0.015],
DisplayFunction->Identity];
Show[fplot,dataplot,AxesLabel->{"x",""},
PlotRange->All,AxesOrigin->{0,0},
DisplayFunction->$DisplayFunction];
```

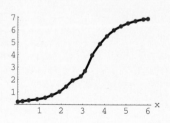

That's a little bit more pleasing than the straight line segments. Here's the resulting calculation of $\int_0^6 f[x]\,dx$:

You can try:

In[27]:=
```
secondestimate = NIntegrate[f[x],{x,0,6},AccuracyGoal->1]
```

Out[27]=
```
19.664
```

Mathematica is telling you that $\int_0^6 f[x]\,dx = 19.664$ within 0.1. Again this is more accuracy than the data points themselves justify. Compare:

In[28]:=
```
{firstestimate,secondestimate}
```

Out[28]=
```
{19.6661, 19.6669}
```

Go with $\int_0^6 f[x]\,dx = 19.6$ and go on to the next job.

■ T.3) Nonsense integrals

When you write $\int_a^b f[x]\,dx$ and attempt to evaluate it, the expression $\int_a^b f[x]\,dx$ is meaningless if:

$\rightarrow$ $f[x]$ is not defined at some x with $a \leq x \leq b$,

or if

$\rightarrow$ $f[x]$ has a singularity (blow up or blow down) at some x with $a \leq x \leq b$.

T.3.a)

Say why neither

$$\int_{-1}^{2}\left(\frac{1}{x^3}\right)dx \qquad \text{nor} \qquad \int_{0}^{1}\left(\frac{1}{x^3}\right)dx$$

has meaning.

Answer: These are meaningless because $1/x^3$ blows up (is unbounded) and blows down on any interval containing 0.

In[29]:=
```
Plot[1/x^3,{x,-0.2,0.2}];
```

$f[x] = 1/x^3$ has a big fat singularity at $x = 0$. *Mathematica* advertises a value of:

In[30]:=
```
Integrate[1/x^3,{x,-1,2}]
```

Out[30]=
$$\frac{3}{8}$$

This answer is wrong because $\int_{-3}^{2} \left(1/x^3\right) dx$ is meaningless.

New versions of *Mathematica* do this correctly.

When you calculate $\int_{-1}^{2} \left(1/x^3\right) dx$ by adding

$$\int_{-1}^{0} \left(\frac{1}{x^3}\right) dx + \int_{0}^{2} \left(\frac{1}{x^3}\right) dx,$$

Mathematica responds correctly:

In[31]:=
```
Integrate[(1/x^3),{x,-1,0}] + Integrate[(1/x^3),{x,0,2}]
```
```
Infinity::indet:
   Indeterminate expression -Infinity + Infinity
      encountered.
```

Out[31]=
```
Indeterminate
```

On the other hand, *Mathematica* correctly declines to answer when you ask for $\int_{0}^{1} \left(1/x^3\right) dx$:

In[32]:=
```
Integrate[1/x^3,{x,0,1}]
```

Out[32]=
```
Infinity
```

T.3.b)

Say why

$$\int_{-1}^{1} \sqrt{x}\, dx$$

is meaningless.

Answer: Within the realm of real numbers, this is meaningless because $\sqrt{x}$ is undefined as a real number for $-1 \leq x < 0$. So there is no physical meaning to the area under the curve $y = \sqrt{x}$ for $-1 \leq x \leq 0$. Let's see what *Mathematica* does:

In[33]:=
```
Clear[x];Integrate[Sqrt[x],{x,-1,1}]
```
Out[33]=
$$\frac{2}{3} + \frac{2\ I}{3}$$

Mathematica here makes the only possible choice and interprets the integral as

$$\int_{-1}^{1} \sqrt{x}\ dx = \int_{-1}^{0} \sqrt{x}\ dx + \int_{0}^{1} \sqrt{x}\ dx$$

$$= \int_{-1}^{0} \sqrt{-1\,(-x)}\ dx + \int_{0}^{1} \sqrt{x}\ dx$$

$$= \int_{-1}^{0} \sqrt{-1}\ \sqrt{-x}\ dx + \int_{0}^{1} \sqrt{x}\ dx$$

$$= \sqrt{-1}\ \int_{-1}^{0} \sqrt{-x}\ dx + \int_{0}^{1} \sqrt{x}\ dx.$$

Then *Mathematica* gives, by a method you will soon understand, that

$$\int_{-1}^{0} \sqrt{-x}\ dx = \frac{2}{3}$$

and

$$\int_{0}^{1} \sqrt{x}\ dx = \frac{2}{3}$$

then comes up with the overall answer:

$$\left(\frac{2}{3}\right) + \left(\frac{2}{3}\right)\sqrt{-1} = \left(\frac{2}{3}\right) + \left(\frac{2}{3}\right) i.$$

T.3.c)   Say why

$$\int_{0}^{1} \sqrt{1 - 4\,x^2}\ dx$$

is meaningless.

Answer: Take a look at a plot:

```
In[34]:=
  Clear[x]
  Plot[Sqrt[1 - 4 x^2],{x,0,1},PlotRange->All];
```

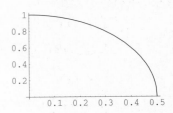

The plot tells you that there is no curve for $0.5 < x \le 1$. No wonder, because

$$1 - 4x^2 < 0 \qquad \text{for} \qquad x > \frac{1}{2}$$

and so $\sqrt{1 - 4x^2}$ is undefined as a real number for $1/2 < x \le 1$. Consequently, $\int_0^1 \sqrt{1 - 4x^2}\,dx$ is meaningless because there is no physical meaning to the area under the curve $y = 1 - 4x^2$ for $1/2 < x \le 1$. See what *Mathematica* does:

```
In[35]:=
  Integrate[Sqrt[1 - 4 x^2],{x,0,1}]
```

```
Out[35]=
  2 I Sqrt[3] + ArcSin[2]
  ───────────────────────
             4
```

Again *Mathematica* was forced into the realm of complex (imaginary) numbers to handle this integral.

T.3.d) | What's the moral?

Answer: Before you try to measure an area, make sure that there is something to measure. If your function $f[x]$ has a singularity between a and b, then $\int_a^b f[x]\,dx$ is meaningless. If your function $f[x]$ is not defined all the way from a to b, then $\int_a^b f[x]\,dx$ is meaningless, but sometimes you can give it meaning as a complex number.

Give It a Try

Experience with the starred ($\star$) problems will be especially beneficial for understanding later lessons.

■ G.1) Plotting and calculating: Symmetry*

G.1.a.i) Here is a plot of $f[x] = x^3$ on $[-2, 2]$ and a little bit of extra information:

```
In[1]:=
  Clear[f,x]; f[x_] = x^3; f[-x] == -f[x]
```

```
Out[1]=
  True
```

```
In[2]:=
  a = 2;
  FilledPlot[f[x],{x,-a,a},
  PlotRange->All,
  AxesLabel->{"x","f[x]"}];
```

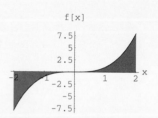

How does the plot give away the value of $\int_{-2}^{2} x^3\, dx$?

G.1.a.ii) Here is a plot of $f[x] = \sin[x]$ on $[-6, 6]$ and a little bit of extra information:

```
In[3]:=
  Clear[f,x]; f[x_] = Sin[x]; f[-x] == -f[x]
Out[3]=
  True
```

```
In[4]:=
  a = 6;
  FilledPlot[f[x],{x,-a,a},
  PlotRange->All,
  AxesLabel->{"x","f[x]"}];
```

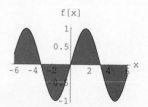

How does the plot give away the value of $\int_{-6}^{6} \sin[x]\, dx$?

G.1.a.iii) Here is a plot of $f[x] = \sin[x^3]$ on $[-2, 2]$ and a little bit of extra information:

```
In[5]:=
  Clear[f,x]; f[x_] = Sin[x^3]; f[-x] == -f[x]
Out[5]=
  True
```

```
In[6]:=
  a = 2;
  FilledPlot[f[x],{x,-a,a},
  PlotRange->All,
  AxesLabel->{"x","f[x]"}];
```

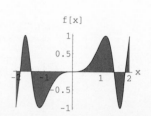

How does the plot give away the value of $\int_{-2}^{2} \sin[x^3]\, dx$?

G.1.a.iv) Here is a plot of $f[x] = x\,e^{-x^2}$ on $[-2.5, 2.5]$ and a little bit of extra information:

In[7]:=
```
Clear[f,x]; f[x_] = x E^(-x^2); f[-x] == -f[x]
```

Out[7]=
True

In[8]:=
```
a = 2.5;
FilledPlot[f[x],{x,-a,a},
PlotRange->All,
AxesLabel->{"x","f[x]"}];
```

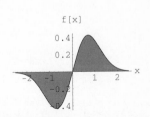

> How does the plot give away the value of $\int_{-2.5}^{2.5} x\,e^{-x^2}\,dx$?

G.1.a.v) Most of the good folks call a function $f[x]$ "odd" if $f[-x] = -f[x]$ for all the x's. All the functions you plotted above are odd functions.

> If $f[x]$ is any odd function and a is any positive constant, then what is the value of $\int_{-a}^{a} f[x]\,dx$ and why?

G.1.b.i) Most good old folks call a function $f[x]$ "even" if $f[-x] = f[x]$ for all the x's. $f[x] = x^2$ and $f[x] = \cos[x]$ are examples of even functions. Take a look:

In[9]:=
```
Clear[f,x]; f[x_] = x^2; f[x] == f[-x]
```

Out[9]=
True

In[10]:=
```
a = 2;
FilledPlot[f[x],{x,-a,a},
AxesLabel->{"x","f[x]"},
AxesOrigin->{0,0}];
```

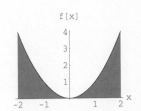

In[11]:=
```
Clear[f,x]; f[x_] = Cos[x]; f[x] == f[-x]
```

Out[11]=
True

In[12]:=
```
a = 6;
FilledPlot[f[x],{x,-a,a},
AxesLabel->{"x","f[x]"},
AxesOrigin->{0,0}];
```

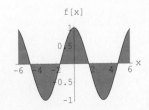

Invent three more examples of even functions and plot them as above.

G.1.b.ii)   Given that $f[x]$ is an even function and a is any positive constant, explain how $\int_{-a}^{a} f[x]\,dx$ is related to $2\int_{0}^{a} f[x]\,dx$.

G.1.c.i)   Go with $f[x] = 5\,x^2 e^{-x}$, $a = 0$, $b = 4$ and look at:

```
In[13]:=
  Clear[f,x]
  f[x_] = 5 x^2 E^(-x);
  a = 0; b = 4;
  firstplot = Plot[f[x],{x,a,b},
  PlotStyle->{{DarkGreen,Thickness[0.02]}},
  AxesLabel->{"x","f[x]"},AxesOrigin->{a,0}];
```

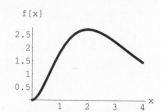

$\int_{a}^{b} f[x]\,dx$ measures the signed area plotted above. Now keep $a = 0$ and $b = 4$ and look at:

```
In[14]:=
  secondplot =
  Plot[f[a + b - x],{x,a,b},
  PlotStyle->{{DarkGreen,Thickness[0.02]}},
  AxesLabel->{"x","f[x]"},AxesOrigin->{b,0}];
```

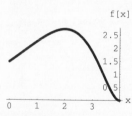

$\int_{a}^{b} f[b + a - x]\,dx$ measures the signed area plotted above.

How do the plots give away the relationship between

$$\int_{a}^{b} f[x]\,dx \qquad \text{and} \qquad \int_{a}^{b} f[b + a - x]\,dx?$$

G.1.c.ii)   Experiment with other functions and other choices of a and b in the style of part G.1.c.i) above.

G.1.c.iii)   Explain in your own terms why you think that the equality

$$\int_{a}^{b} f[x]\,dx = \int_{a}^{b} f[b + a - x]\,dx$$

holds up for any function $f[x]$ and any choice of a and b with $a < b$.

■ G.2) More plotting and integrating*

G.2.a) Here is a plot of $f[x] = x(x-1)(x-2)$ for $0 \leq x \leq 2$:

```
In[15]:=
    Clear[f,x]
    f[x_] = x (x - 1) (x - 2);
    a = 0; b = 2;
    FilledPlot[f[x],{x,a,b},
    AxesLabel->{"x","f[x]"},
    AxesOrigin->{a,0}];
```

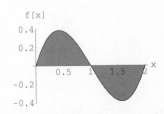

> Use this plot to help you determine the value of $\int_0^2 x(x-1)(x-2)\,dx$.

G.2.b)
> Plot $f[x] = \sin[x]$ for $0 \leq x \leq 2\pi$. Use your plot to help you determine the value of $\int_0^{2\pi} \sin[x]\,dx$.

G.2.c)
> Plot $f[x] = \cos[x]$ for $0 \leq x \leq 2\pi$. Use your plot to help you determine the value of $\int_0^{2\pi} \cos[x]\,dx$.

G.2.d.i) Here are plots of $\sin[x]^2$ and $\cos[x]^2$ on the same axes for $0 \leq x \leq \pi$:

```
In[16]:=
    Clear[f,g,x]; a = 0; b = Pi;
    f[x_] = Sin[x]^2; g[x_] = Cos[x]^2;
    Plot[{f[x],g[x]},{x,a,b},
    PlotStyle->{{Thickness[0.01]},
    {Red,Thickness[0.01]}},
    AxesLabel->{"x",""},AxesOrigin->{a,0}];
```

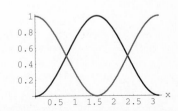

> Use this plot to help you determine a relationship between $\int_0^\pi \sin[x]^2\,dx$ and $\int_0^\pi \cos[x]^2\,dx$.

G.2.d.ii) Here are plots of $\sin[x]^2$ and $\cos[x]^2$ on the same axes for $0 \leq x \leq 2\pi$:

```
In[17]:=
    Clear[f,g,x];
    f[x_] = Sin[x]^2; g[x_] = Cos[x]^2;
    a = 0; b = 2 Pi;
    Plot[{f[x],g[x]},{x,a,b},
    PlotStyle->{{Thickness[0.01]},
    {Red,Thickness[0.01]}},
    AxesLabel->{"x",""},AxesOrigin->{a,0}];
```

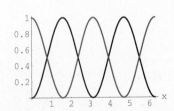

Use this plot to help you determine a relationship between $\int_0^{2\pi} \sin[x]^2\, dx$ and $\int_0^{2\pi} \cos[x]^2\, dx$.

■ G.3) Calculating some integrals*

G.3.a) Look at this plot of $f[x] = 2x - 3$ for $3 \le x \le 5$.

```
In[18]:=
  Clear[f,x]
  f[x_] = 2 x - 3;
  Plot[f[x],{x,3,5},
  PlotStyle->{{Thickness[0.01],DarkGreen}},
  PlotRange->{0,f[5]},
  AxesLabel->{"x","f[x]"}];
```

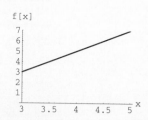

Use what you know about areas inside trapezoids to calculate the value of $\int_3^5 f[x]\, dx$. Check with *Mathematica*.

G.3.b.i) Look at this plot of $f[x] = 1 - x$ for $-1 \le x \le 1$.

```
In[19]:=
  Clear[f,x]
  f[x_] = 1 - x;
  Plot[f[x],{x,-1,1},
  PlotStyle->{{Thickness[0.01],Blue}},
  PlotRange->All,
  AxesLabel->{"x","f[x]"}];
```

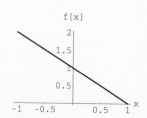

Use what you know about areas inside triangles to calculate the value of $\int_{-1}^1 f[x]\, dx$. Check with *Mathematica*.

G.3.b.ii) Look at this plot of $f[x] = 1 - x$ for $-1 \le x \le 2$.

```
In[20]:=
  Clear[f,x]
  f[x_] = 1 - x;
  Plot[f[x],{x,-1,2},
  PlotStyle->{{Thickness[0.01],Blue}},
  PlotRange->All,
  AxesLabel->{"x","f[x]"}];
```

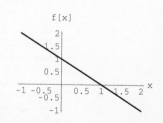

Use what you know about areas inside triangles to calculate the value of $\int_{-1}^{2} f[x]\, dx$. Check with *Mathematica*.

G.3.c) Look at this plot of $f[x] = \sqrt{16 - x^2}$ for $0 \le x \le 4$.

```
In[21]:=
  Clear[f,x]
  f[x_] = Sqrt[16 - x^2];
  Plot[f[x],{x,0,4},
  PlotStyle->{{Thickness[0.01],Blue}},
  AspectRatio->Automatic,PlotRange->All,
  AxesLabel->{"x","f[x]"}];
```

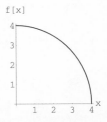

Use what you know about areas inside circles to calculate the value of

$$\int_{0}^{4} \sqrt{16 - x^2}\, dx.$$

Check with *Mathematica*.

G.3.d) Look at this plot of $f[x] = x\sqrt{1 - x^2}$ for $-1 \le x \le 1$.

```
In[22]:=
  Clear[f,x]
  f[x_] = x Sqrt[1 - x^2];
  Plot[f[x],{x,-1,1},
  PlotStyle->{{Thickness[0.01],DarkGreen}},
  AxesLabel->{"x","f[x]"}];
```

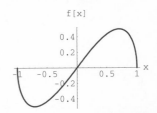

Use this plot to read off the value of $\int_{-1}^{1} x\sqrt{1 - x^2}\, dx$.

G.3.e) Look at this plot of $f[x] = x$ for $-1 \le x \le 1$.

```
In[23]:=
  Clear[f,x]; f[x_] = x;
  Plot[f[x],{x,-1,1},
  PlotStyle->{{Thickness[0.01],Blue}},
  AspectRatio->Automatic,PlotRange->All,
  AxesLabel->{"x","f[x]"}];
```

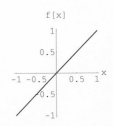

Look at this plot of $g[x] = |x|$ for $-1 \le x \le 1$.

In[24]:=
```
Clear[g,x]
g[x_] = Abs[x]; Plot[g[x],{x,-1,1},
PlotStyle->{{Thickness[0.01],Blue}},
AspectRatio->Automatic,PlotRange->All,
AxesLabel->{"x","f[x]"}];
```

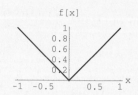

Use whatever methods you like to calculate $|\int_{-1}^{1} x\,dx|$ and $\int_{-1}^{1} |x|\,dx$. Try to explain the results in terms of the plots.

■ G.4) Experiments geared toward breaking the code of the integral⋆

Experienced integral watchers know how to break the code of the integral. The first step toward breaking the code of the integral is to learn how to calculate $f'[t]$ when $f[t]$ is given by $f[t] = \int_{a}^{t} g[x]\,dx$ for some other function $g[x]$.

G.4.a) Go with the specific case of $f[t] = \int_{a}^{t} g[x]\,dx$ with $g[x] = \cos[3\,x] + 1/2$ and $a = 0$. Here's a plot of $g[x]$ for $a \leq x \leq b = 4$:

In[25]:=
```
Clear[f,g,x,t]; a = 0; b = 4;
g[x_] = Cos[3 x] + 1/2;
gplot = Plot[g[x],{x,a,b},
PlotStyle->{{Red,Thickness[0.01]}},
AxesLabel->{"x","g[x]"}];
```

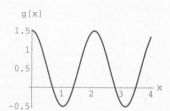

And here comes a plot of $f[t] = \int_{a}^{t} g[x]\,dx$ for $a \leq x \leq b$:

In[26]:=
```
f[t_] := NIntegrate[g[x],{x,a,t},AccuracyGoal->2];
fplot = Plot[f[t],{t,a,b},
PlotStyle->{{Blue,Thickness[0.02]}},
DisplayFunction->Identity];
Show[fplot,AxesLabel->{"x","f[x]"},
DisplayFunction->$DisplayFunction];
```

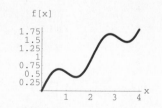

Here they are together:

In[27]:=
```
Show[fplot,gplot,
DisplayFunction->$DisplayFunction];
```

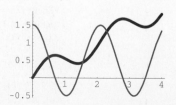

Describe what you see, paying special attention to what $f[t] = \int_a^t g[x]\,dx$ (thick) is doing when $g[t]$ (thin) is positive and to what $f[t] = \int_a^t g[x]\,dx$ is doing when $g[t]$ is negative.

What clue does this give you about the relationship between $f'[t]$ and $g[t]$?

G.4.b) Go with the specific case of

$$f[t] = \int_a^t g[x]\,dx \qquad \text{with } g[x] = x \sin[2\,x] + \frac{x}{2} \qquad \text{and } a = -2.$$

Here's a plot of $g[x]$ for $a \le x \le b = 4$:

In[28]:=
```
a = -2; b = 4;
Clear[f,g,x,t]
g[x_] = x Sin[2 x]+ x/2;
gplot = Plot[g[x],{x,a,b},
PlotStyle->{{Red,Thickness[0.01]}},
AxesLabel->{"x","g[x]"}];
```

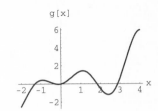

And here comes a plot of $f[t] = \int_a^t g[x]\,dx$ for $a \le x \le b$:

In[29]:=
```
f[t_] := NIntegrate[g[x],{x,a,t},AccuracyGoal->2];
fplot = Plot[f[t],{t,a,b},
PlotStyle->{{Blue,Thickness[0.02]}},
DisplayFunction->Identity];
Show[fplot,AxesLabel->{"x","f[x]"},
DisplayFunction->$DisplayFunction];
```

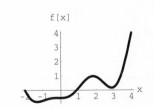

Here they are together:

In[30]:=
```
Show[fplot,gplot,
DisplayFunction->$DisplayFunction];
```

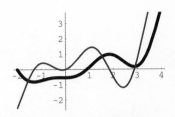

Describe what you see, paying special attention to what $f[t] = \int_a^t g[x]\,dx$ (thick) is doing when $g[t]$ (thin) is positive and to what $f[t] = \int_a^t g[x]\,dx$ is doing when $g[t]$ is negative.

What clue does this give you about the relationship between $f'[t]$ and $g[t]$?

G.4.c) Make up your own function $g[x]$ and your own choices of a and b and run the same experiment. Try to use a function $g[x]$ that has both positive and negative values for different x's with $a \leq x \leq b$.

Describe what you see, paying special attention to what $f[t] = \int_a^t g[x]\, dx$ is doing when $g[t]$ is positive and to what $f[t] = \int_a^t g[x]\, dx$ is doing when $g[t]$ is negative.

What clue does this give you about the relationship between $f'[t]$ and $g[t]$?

G.4.d) Repeat part G.4.c) with new choices of $g[x]$, a, and b.

G.4.e.i) This time go with the specific case of $f[t] = \int_a^t g[x]\, dx$ with $g[x] = x^2$ and $a = 1$. But this time let *Mathematica* get the formula for $f[t]$:

In[31]:=
```
Clear[f,g,x,t]; a = 1; g[x_] = x^2;
f[t_] = Integrate[g[x],{x,a,t}]
```

Out[31]=

$$-\left(\frac{1}{3}\right) + \frac{t^3}{3}$$

Now let *Mathematica* calculate $f'[t]$:

In[32]:=
```
f'[t]
```

Out[32]=

$$t^2$$

How does this calculation of $f'[t]$ square with the outcomes of your experiments above?

G.4.e.ii) Repeat the experiment in part G.4.e.i) immediately above for each of the choices of $g[x]$ and a used in parts a), b), c), and d) above.

Comment on the outcomes.

■ G.5) Using numerical integration⋆

G.5.a.i) Use NIntegrate to estimate the value of

$$4 \int_0^{1/\sqrt{2}} \frac{1}{\sqrt{1 - x^2}}\, dx.$$

G.5.a.ii) Here is an attempt to use NIntegrate to estimate the value of

$$\int_0^2 \frac{1}{1 - x^2}\, dx :$$

In[33]:=
```
Clear[x]
NIntegrate[1/(1 - x^2),{x,0,2}]
```
```
                                  1
Power::infy: Infinite expression - encountered.
                                  0
```
```
NIntegrate::inum:
   Integrand ComplexInfinity is not numerical at {x} = {1}.
```

Out[33]=
```
             1
NIntegrate[------, {x, 0, 2}]
                2
            1 - x
```

> What is it about the function
>
> $$\frac{1}{1 - x^2}$$
>
> that drove NIntegrate bats?

G.5.b) The formula $f[x] = 0.5\, e^{2\sin[3.14x]}$ looks innocent enough. Here's how it plots out:

In[34]:=
```
Clear[f,x]
f[x_] = 0.5 E^(2 Sin[3.14 x]);
Plot[f[x],{x,-2,2},
PlotStyle->{{Thickness[0.01],DarkGreen}},
PlotRange->All,AxesLabel->{"x","f[x]"}];
```

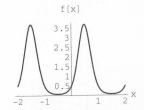

> Use NIntegrate to get an estimate of $\int_{-2}^{2} f[x]\, dx$.

G.5.c) Plot enough of

$$y^2 = \frac{x^2\,(3 + x)}{3 - x}$$

to discover the loop. Then use NIntegrate to calculate the area the loop encloses.

■ G.6) The flavor of NIntegrate⋆

This problem appears only in the electronic version.

■ G.7) Estimating the area of a piece of ground

Calculus&*Mathematica* thanks Illinois farmer Tim Taylor for allowing the use of his plat book as a basis for this problem.

G.7.a) Illinois farmer Tim Taylor is negotiating a contract to cash farm a piece of ground along the Salt Fork River. Needing an estimate of the acreage involved and not having time to wait for a surveying crew, he goes out in the field and makes some measurements along the boundary of the field. Here is a picture reporting his measurements (all in feet):

In[35]:=
```
boundarypoints =
{{0,1712},{10,1792},{21,1872},{30,1948},{40,2031},{50,2114},{60,2193},{71,2264},
 {80,2336},{90,2420},{100,2492},{109,2565},{120,2645},{130,2708},{139,2787},
 {150,2855},{158,2914},{170,2967},{180,3034},{193,3089},{200,3127},{210,3183},
 {222,3218},{230,3251},{243,3278},{250,3298},{260, 3314},{271, 3325},{280,3307},
 {290,3295},{302,3284},{310,3242},{324, 3205},{330,3144},{340,3074},{350,2994},
 {360,2890},{370,2765},{380,2636},{390,2494},{400,2328},{410,2141},{423,1945},
 {431,1724},{442, 1475},{448,1198},{460,906},{471, 576},{480, 230},{486,0}};
```

In[36]:=
```
pointplot = ListPlot[boundarypoints,PlotStyle->
{Red,PointSize[0.01]},DisplayFunction->Identity];
boundaryplot = {Graphics[{Blue,Line[boundarypoints]}],
Graphics[{Thickness[0.01],Blue,
Line[{{0,1712},{0,0},{486,0}}]}]};
Show[pointplot,boundaryplot,AxesLabel->{"x","y"},
PlotRange->All,AxesOrigin->{0,0},
DisplayFunction->$DisplayFunction];
```

That's the riverbank on the top and right together with the typical Illinois straight boundary segments at the left and bottom. Tim thinks: "I can get out my calculator and get an estimate of the total acreage, but that's going to take a lot of finger punching and I've got better things to do with my time." Knowing that you have access to *Mathematica*, he calls you up and asks you to carry out the calculation on *Mathematica*. You say: "Sure, but first tell me how many square feet are in one acre." Tim replies that there are 43,560 square feet in one acre.

> Give Tim his estimate.

■ G.8) Integration of data lists

Sometimes the actual formula for a function is not available, but a reasonably complete tabulation of its values is available. This situation is common when a laboratory experiment is over and the data are organized.

G.8.a) The following values $\{x, f[x]\}$ are known:

In[37]:=
```
tabulation =
  {{0.00,0.00},{0.621,0.215},{0.947,0.610},{1.603,0.929},{1.959,1.099},
   {2.634,1.164},{3.044,1.297},{3.648,1.389},{4.057,1.602},{4.624,1.794},
   {5.075,1.880},{5.534,1.967},{5.893,2.116},{6.522,2.169},{6.931,2.272},
   {7.381,2.365},{7.892,2.448},{8.593,2.492},{9.187,2.506}};
```

In[38]:=
```
pointplot =
ListPlot[tabulation,
PlotStyle->{Red,PointSize[0.02]}];
```

But the exact formula for $f[x]$ is not known.

What is your best shot at a value for $\int_0^{9.187} f[x]\,dx$?

G.8.b) The following values $\{x, f[x]\}$ are known:

In[39]:=
```
tabulation =
  {{0.00,13.376},{0.402,8.551},{0.819,1.611},{1.188,-3.969},{1.407,-6.160},
   {1.944,-5.108},{2.094,-2.227},{2.409,0.677},{2.701,2.566},{2.996,2.749},
   {3.673,1.804},{3.780,0.363},{4.107,-0.847},{4.537,-1.318},{4.89,-1.023},
   {5.401,-0.363},{5.632,0.202},{5.962,0.554},{6.117,0.576},{6.470,0.410},
   {7.00,0.007}};
```

In[40]:=
```
pointplot =
ListPlot[tabulation,
PlotStyle->{Red,PointSize[0.02]}];
```

But the exact formula for $f[x]$ is not known.

What is your best shot at a value for $\int_0^7 f[x]\,dx$?

■ G.9) Integrating, plotting, measuring, and guessing

G.9.a.i) For $t \geq 0$, set $f[t] = \int_0^t \cos[x]\,dx$. Here is a plot of $f[t]$:

```
In[41]:=
    Clear[f,t,x]
    f[t_] := NIntegrate[Cos[x],
    {x,0,t},AccuracyGoal->2];
    fplot = Plot[f[t],{t,0,2 Pi},
    PlotStyle->{{Thickness[0.01],Blue}},
    AxesLabel->{"t","f[t]"}];
```

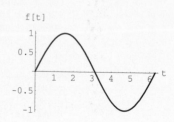

Here's the value of $f[\pi/2]$:

```
In[42]:=
    f[Pi/2]

Out[42]=
    1.
```

> Plot the area that $f[\pi/2]$ measures.

G.9.a.ii) For $t \geq 0$, again set $f[t] = \int_0^t \cos[x]\,dx$. Here's another look at the plot of $f[t]$:

```
In[43]:=
    Show[fplot];
```

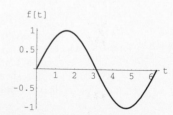

> Use the plot to guess a formula for $f[t]$. Check yourself with a plot.

G.9.b.i) For $t \geq 0$, set $f[t] = \int_0^t \sin[x]\,dx$. Here is a plot of $f[t]$:

```
In[44]:=
    Clear[f,t,x]
    f[t_] := NIntegrate[Sin[x],{x,0,t},AccuracyGoal->2];
```

```
In[45]:=
    fplot =
    Plot[f[t],{t,0,2 Pi},
    PlotStyle->{{Thickness[0.01],Blue}},
    AxesLabel->{"t","f[t]"}];
```

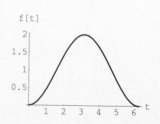

Here is the value of $f[\pi]$:

In[46]:=
f[Pi]

Out[46]=
2.

Plot the area that $f[\pi]$ measures.

G.9.b.ii) For $t \geq 0$, again set $f[t] = \int_0^t \sin[x]\, dx$. Here's another look at the plot of $f[t]$:

In[47]:=
Show[fplot];

Use the plot to guess a formula for $f[t]$. Check yourself with a plot.

LESSON 2.02

Breaking the Code of the Integral: The Fundamental Formula

Basics

■ **B.1)** If $f[t]$ is given by $f[t] = \int_a^t g[x]\,dx$, then $f'[t] = g[t]$

Experienced integral watchers know how to break the code of the integral. The first step toward this is to learn how to calculate $f'[t]$ when $f[t]$ is given by

$$f[t] = \int_a^t g[x]\,dx$$

for some other function $g[x]$. Go with the specific case of $f[t] = \int_a^t g[x]\,dx$ with $g[x] = x\sin[3\,x] + 2$ and $a = 1$. Here's a plot of $g[x]$ for $a \le x \le b = 5$:

In[1]:=
```
Clear[f,g,x,t]; a = 1; b = 5;
g[x_] = x Sin[3 x] + 2;
gplot = Plot[g[x],{x,a,b},
PlotStyle->{{Blue,Thickness[0.02]}},
AxesLabel->{"x","g[x]"}];
```

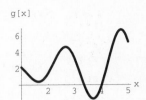

And here comes a plot of $f[t] = \int_a^t g[x]\,dx$ for $a \le t \le b$:

In[2]:=
```
f[t_] := NIntegrate[g[x],{x,a,t},AccuracyGoal->2];
fplot = Plot[f[t],{t,a,b},
PlotStyle->{{Blue,Thickness[0.02]}},
AxesLabel->{"t","f[t]"},
Epilog->Text[FontForm[" f[t] = # g[x] dx ",
{"CalcMath",7}],3,7}]];
```

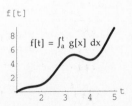

41

Put in trapezoids that approximate $\int_a^b g[x]\,dx$:

In[3]:=
```
Clear[points,pointplot,trapezoid,iterations,jump,trapezoids,trapplot];
jump[iterations_] = (b - a)/iterations;

points[iterations_] := Table[{x,g[x]},{x,a,b,jump[iterations]}];

pointplot[iterations_] := ListPlot[points[iterations],
PlotStyle->{Red,PointSize[0.03]},DisplayFunction->Identity];

Clear[trapezoid,x]
trapezoid[x_,iterations_] := Graphics[{Thickness[0.01],Line[{{x,0},
{x + jump[iterations],0},{x + jump[iterations],g[x + jump[iterations]]},
{x,g[x]},{x,0}}]}];

trapezoids[iterations_] := Table[trapezoid[x,iterations],{x,a,b - jump[iterations],
jump[iterations]}];

trapplot[iterations_] := Show[gplot,pointplot[iterations],trapezoids[iterations],
AxesLabel->{"x",""},AxesOrigin->{a,0},PlotLabel->"g[x] and trapezoids",
AxesLabel->{"x","g[x]"},PlotRange->All,DisplayFunction->Identity];
```

In[4]:=
```
Show[GraphicsArray[{{fplot},{trapplot[8]}}],
DisplayFunction->$DisplayFunction];
```

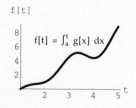

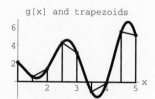

Now look at this:

In[5]:=
```
start = {a,0};
Clear[jump,iterations]; jump[iterations_] = (b - a)/iterations;

Clear[t,y,next,point,k]; next[{t_,y_},iterations_] :=
{t,y} + jump[iterations]{1,(g[t] + g[t + jump[iterations]])/2};

point[0,iterations_] = start; point[k_,iterations_] :=
point[k,iterations] = N[next[point[k - 1,iterations],iterations]];

Clear[Trappoints,Trappointplot,Trapareaplotter,Trapareacurve]
Trappoints[iterations_] := Table[point[k,iterations],{k,0,iterations}];

Trappointplot[iterations_] := ListPlot[Trappoints[iterations],
PlotStyle->{Red,PointSize[0.03]},DisplayFunction->Identity];
```

```
Trapareacurve[iterations_] := Graphics[{Thickness[0.01],Red,
Line[Trappoints[iterations]]}];

Trapareaplotter[iterations_] := Show[fplot,Trappointplot[iterations],
Trapareacurve[iterations], AxesLabel->{"x",""},
PlotRange->All,PlotLabel-> "f[t] and accum. area of 'zoids ",
DisplayFunction->Identity];
```

In[6]:=
```
Show[GraphicsArray[{{Trapareaplotter[8]},{trapplot[8]}}],
DisplayFunction->$DisplayFunction];
```

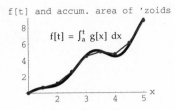

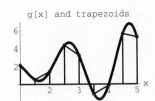

As you go from left to right, the height of each plotted point on the upper plot measures the accumulated area inside the trapezoids that fall to the left of the corresponding point on the lower plot. This is why the broken line curve in the lower plot has no choice but to be a reasonable approximation of the plot of $f[t] = \int_a^t g[x]\,dx$. See what happens when you go from low iteration numbers (big jumps) to high iteration numbers (tiny jumps):

In[7]:=
```
Clear[k]
Table[Show[GraphicsArray[{{Trapareaplotter[2^k]},{trapplot[2^k]}}],
DisplayFunction->$DisplayFunction];,{k,2,5}];
```

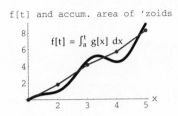

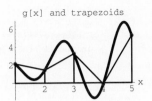

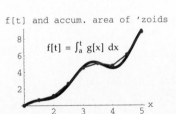

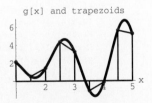

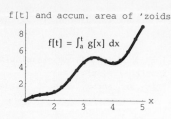

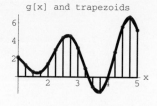

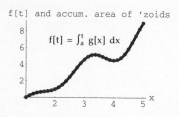

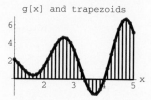

As the iteration number increases, the horizontal jump between the points decreases and the broken line approximation of the $f[t] = \int_a^t g[x]\,dx$ curve (left side) improves in direct proportion to the quality of the trapezoidal approximation of $\int_a^b f[x]\,dx$ (right side).

B.1.a) Try it for a different function $g[x]$ and another choice of endpoints a and b.

Discuss the outcome.

Answer: Copy, paste, and edit. Just for the heck of it, go with the specific case of $f[t] = \int_a^t g[x]\,dx$ with $g[x] = \sin[x^2]$ and $a = -1$. Here's a plot of $g[x]$ for $a \le x \le b = 3$:

```
In[8]:=
  a = -1; b = 3; Clear[f,g,x,t];
  g[x_] = Sin[x^2];
  gplot = Plot[g[x],{x,a,b},
  PlotStyle->{{Blue,Thickness[0.02]}},
  AxesLabel->{"x","g[x]"}];
```

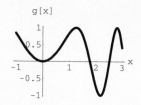

And here comes a plot of $f[t] = \int_a^t g[x]\,dx$ for $a \le x \le b$:

```
In[9]:=
  f[t_] := NIntegrate[g[x],
  {x,a,t},AccuracyGoal->2]; fplot = Plot[f[t],{t,a,b},
  PlotStyle->{{Blue,Thickness[0.02]}},
  AxesLabel->{"t","f[t]"},
  Epilog->Text[FontForm[" f[t] = #f g[x] dx ",
  {"CalcMath",7}],{1.75,0.6}]];
```

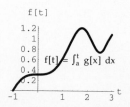

Now watch the action:

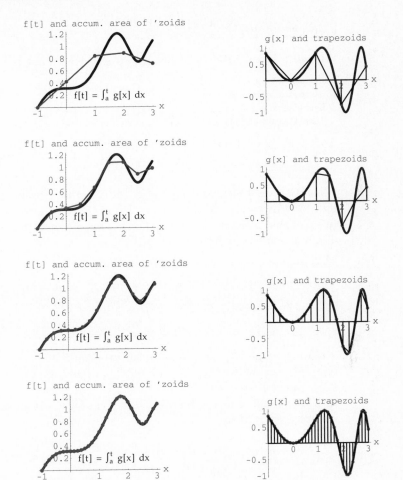

Again, as you go down, the height of each plotted point on the left plot measures the accumulated area inside the trapezoids that fall to the left of the corresponding point on the right plot. This is why the broken line curve in the lower plot has no choice but to be a reasonable approximation to the plot of $f[t] = \int_a^t g[x]\,dx$. See what happens when you go from low iteration numbers (big jumps) to high iteration numbers (tiny jumps). Again, as the iteration number increases, the horizontal jump between the points decreases and the broken line approximation of $f[t] = \int_a^t g[x]\,dx$ curve (left plots) improves in direct proportion to the quality of the trapezoidal approximation of $\int_a^b f[x]\,dx$ (right plots). This will happen for any function $g[x]$ and any endpoints a and b you go with because when you make the jumps small enough, the accumulated area measurements of the trapezoids give a good approximation to $\int_a^b g[x]\,dx$.

B.1.b) How do the plots and explanations above confirm the following rule once and for all: If you go with a function $g[x]$ and you put $f[t] = \int_a^t g[x]\,dx$, then you can be certain that $f'[t] = g[t]$?

Answer: Here's the code that measures the accumulated area of the trapezoids:

In[10]:=
```
Clear[t,y,next,point,k]
next[{t_,y_},iterations_] := {t,y} +
jump[iterations]{1,(g[t] + g[t + jump[iterations]])/2};
```

This code moves from the point $\{t, y\}$ with the growth rate

$$\frac{g[t] + g[t + \text{jump}]}{2}.$$

When the iteration number is high, the jump is tiny and this number is close to $g[t]$. At this point, the Race Track Principle steps in to tell you that the broken line plot that measures the accumulated area of the trapezoids actually fakes a plot of a function whose derivative is $g[t]$.

Math happens again, and it's all in your favor.

B.1.c) Put $f[t] = \int_{1.03}^t e^{-x} \cos[3\,x^4]\,dx$, calculate $f'[t]$, and discuss the big practical advantage you got.

Answer: One way to try to do this problem is to attempt to calculate $f[t]$ and then take the derivative. The only hitch is that neither *Mathematica* nor anything else can come up with a neat formula for $f[t]$. The practical advantage you get is: This doesn't hold you back because you can do the calculation of $f'[t]$ by hand instantly. Start with

$$f[t] = \int_{1.03}^t e^{-x} \cos[3\,x^4]\,dx.$$

This gives you

$$f'[t] = e^{-t} \cos[3\,t^4].$$

Done. That's all there is to it. No chain rule to worry about; no nothing to worry about.

Fat City.

■ B.2) The fundamental formula $f[t] - f[a] = \int_a^t f'[x]\,dx$

B.2.a) The fundamental formula of calculus completely breaks the code of the integral by telling you that $f[t] - f[a] = \int_a^t f'[x]\,dx$.

Where does this beauty come from?

Answer: The Race Track Principle strikes again. Fix a point a and go to the race track and run a race between two functions:

$$\text{Newton}[t] = f[t] - f[a]$$

and

$$\text{Leibniz}[t] = \int_a^t f'[x]\, dx.$$

When $t = a$, you get

$$\text{Newton}[a] = f[a] - f[a] = 0$$

and

$$\text{Leibniz}[a] = \int_a^a f'[x]\, dx = 0$$

because $\int_a^a f'[x]\, dx$ measures the area of a (degenerate) rectangle whose height is $f'[a]$ and whose base is 0. This means that if you start the race at $t = a$, the two functions Leibniz$[t]$ and Newton$[t]$ start their race together. If you can see why

$$\text{Newton}'[t] = \text{Leibniz}'[t] \qquad \text{for } t > a,$$

then the Race Track Principle will step in to tell you that they remain tied throughout their race; that is,

$$\text{Newton}[t] = \text{Leibniz}[t] \qquad \text{for all } t\text{'s with } t > a.$$

To see why Newton$'[t] = $ Leibniz$'[t]$ for $t > a$, note Newton$[t] = f[t] - f[a]$; so

$$\text{Newton}'[t] = f'[t] - 0 = f'[t]$$

(because a is a constant) and

$$\text{Leibniz}[t] = \int_a^t f'[x]\,dx;$$

so

$$\text{Leibniz}'[t] = f'[t]$$

((B.1) above.) The upshot: Newton$'[t] = $ Leibniz$'[t]$. Now you know that

$$\text{Newton}[t] = \text{Leibniz}[t].$$

This is the same as

$$f[t] - f[a] = \int_a^t f'[x]\, dx.$$

And you're out of here.

B.2.b) Use the fundamental formula to calculate the exact value of $\int_0^{\pi/2} \sin[x]\, dx$ and explain what the calculation measures.

Answer: The fundamental formula says

$$f[t] - f[a] = \int_a^t f'[x]\, dx.$$

To use the fundamental formula to calculate $\int_0^{\pi/2} \sin[x]\,dx$, look for a function $f[x]$ with

$$f'[x] = \sin[x].$$

A moment's reflection tells you that

$$f[x] = -\cos[x]$$

will do:

In[11]:=
```
Clear[x,f]; f[x_] = -Cos[x]; f'[x]
```
Out[11]=
```
Sin[x]
```

Good. The fundamental formula,

$$f\left[\frac{\pi}{2}\right] - f[0] = \int_0^{\pi/2} f'[x]\,dx = \int_0^{\pi/2} \sin[x]\,dx,$$

tells you that $\int_0^{\pi/2} \sin[x]\,dx$ is given by:

In[12]:=
```
f[Pi/2] - f[0]
```
Out[12]=
```
1
```

Check:

In[13]:=
```
Integrate[Sin[x],{x,0,Pi/2}]
```
Out[13]=
```
1
```

Nailed it. The calculation $\int_0^{\pi/2} \sin[x]\,dx = 1$ means that the shaded area below measures out to one square unit:

In[14]:=
```
FilledPlot[Sin[x],{x,0,Pi/2},
AspectRatio->Automatic,
AxesLabel->{"x","y"},
PlotStyle->Blue,
PlotLabel->"y = Sin[x]"];
```

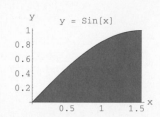

Think of it. That shaded area measures out to exactly one square unit; no more, no less.

B.2.c)   What calculational advantage does the fundamental formula give you?

Answer: Without resorting to approximations, you can calculate the exact value of an integral $\int_a^t g[x]\,dx$ if you can come up with a function $f[x]$ with $f'[x] = g[x]$. Once you have your hands on $f[x]$, you just write

$$\int_a^t g[x]\,dx = f[t] - f[a]$$

and you're done.

B.2.d) | What calculational advantage does the fundamental formula give *Mathematica*?

Answer: *Mathematica* uses the fundamental formula just the way you use it. When you ask *Mathematica* to calculate the exact value of $\int_a^t g[x]\,dx$, *Mathematica* looks for a function $f[x]$ with $f'[x] = g[x]$. Once *Mathematica* has its hands on $f[x]$, *Mathematica* reports $\int_a^t g[x]\,dx = f[t] - f[a]$. Watch *Mathematica* use the fundamental formula:

In[15]:=
```
Clear[x,g,t]; g[x_] = x E^(-x);
Integrate[g[x],{x,2,t}]
```

Out[15]=
$$\frac{3}{E^2} + \frac{-1-t}{E^t}$$

In[16]:=
```
Clear[x,g,t]; g[x_] = Sin[x/3];
Integrate[g[x],{x,Pi,t}]
```

Out[16]=
$$\frac{3}{2} - 3\,\mathrm{Cos}\!\left[\frac{t}{3}\right]$$

The fundamental formula is directly programmed into *Mathematica*:

In[17]:=
```
Clear[f,x,a,t,Derivative]
Integrate[f'[x],{x,a,t}]
```

Out[17]=
```
-f[a] + f[t]
```

The fundamental formula is a big deal.

■ B.3) Measurements of distance and velocity via the fundamental formula

If an object moves on a straight line and is $f[t]$ units away from a reference marker at time t, then its velocity is given by $\mathrm{vel}[t] = f'[t]$ and its acceleration is given by $\mathrm{accel}[t] = \mathrm{vel}'[t]$.

The fundamental formula says

$$f[t] - f[a] = \int_a^t f'[x]\, dx = \int_a^t \text{vel}[x]\, dx;$$

so

$$f[t] = f[a] + \int_a^t \text{vel}[x]\, dx.$$

And the fundamental formula says

$$\text{vel}[t] - \text{vel}[a] = \int_a^t \text{vel}'[x]\, dx = \int_a^t \text{accel}[x]\, dx;$$

so

$$\text{vel}[t] = \text{vel}[a] + \int_a^t \text{accel}[x]\, dx.$$

The fundamental formula has just told you that if you know vel[t], then you can calculate $f[t]$; and if you know accel[t], then you can calculate both vel[t] and $f[t]$. No more verbiage now; it's time to put this good stuff into action.

B.3.a) Here's a line and a reference marker:

In[18]:=
```
line = Graphics[{Thickness[0.02],
Line[{{0,0},{50,0}}]}];
marker = Graphics[{Red,Thickness[0.01],
Line[{{0.5,0.5},{0.5,-0.5}}]}];
Show[line,marker,PlotRange->{-3,3},
AspectRatio->1/4];
```

An object moves from left to right on this line. When you turn on the timer $(t = 0)$, the object is 5.7 units to the right of the marker. And t seconds after you turn on the timer, the object is moving to the right with velocity

$$\text{vel}[t] = 1.2\, t + 2.3\, e^{-t} \text{ units per second.}$$

Put

$$f[t] = \text{distance of the object from the right of the marker } t \text{ seconds}$$
$$\text{after you turn on the timer.}$$

Use the fundamental formula to give a formula for $f[t]$.

Then plot $f[t]$ for $0 \le t \le 8$ and report how far to the right of the marker the object is when $t = 8$.

Make a little movie portraying the motion of the object.

Answer: Here's what you know:

$\rightarrow$ $f[0] = 5.7$ because that's where the object is when $t = 0$.

■ B.4) $\int_a^\infty f[x]\,dx$ and the fundamental formula

B.4.a) How do you make sense of an integral like $\int_1^\infty e^{-x} dx$?

Answer: You can't say that $\int_1^\infty e^{-x} dx$ measures a concrete area, because the horizontal measurement would be infinite. So you do the next best thing by agreeing that $\int_1^\infty e^{-x} dx$ is the limiting case of

$$\int_1^t e^{-x} dx \qquad \text{as } t \to \infty.$$

In short,

$$\int_1^\infty e^{-x} dx = \lim_{t\to\infty} \int_1^t e^{-x} dx.$$

Some folks regard this as a cop-out and call an integral like $\int_1^\infty e^{-x} dx$ an "improper integral." But this terminology, although in common use, is unnecessarily disparaging.

Now the calculation: To calculate $\int_1^\infty e^{-x} dx$, you let *Mathematica* use the fundamental formula to get:

In[33]:=
```
Clear[x,t]
Integrate[E^(-x),{x,1,t}]
```
Out[33]=
```
1    -t
- - E
E
```

This tells you that

$$\int_1^t e^{-x} dx = \frac{1}{e} - e^{-t}.$$

So

$$\int_1^\infty e^{-x} dx = \lim_{t\to\infty} \int_1^t e^{-x} dx = \lim_{t\to\infty} \frac{1}{e} - e^{-t} = \frac{1}{e}.$$

Check:

In[34]:=
```
Integrate[E^(-x),{x,1,Infinity}]
```
Out[34]=
```
1
-
E
```

Got it. You might ask how $\int_1^\infty e^{-x} dx$ could be finite. If this question bothers you, then look at the plot of e^{-x}:

$In[35]:=$
```
Plot[E^(-x),{x,1,100},
PlotRange->All,
PlotStyle->
 {{Red,Thickness[0.01]}}];
```

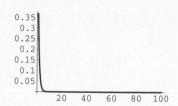

The reason $\int_1^\infty e^{-x}dx$ is finite is that most of the area under the e^{-x} curve for $1 \le x < \infty$ is concentrated over the interval $1 \le x \le 10$. For $x \ge 10$ very little extra area comes in. To confirm this, look at the following decimal approximations of $\int_1^\infty e^{-x}dx$ and $\int_1^{10} e^{-x}dx$:

$In[36]:=$
```
{N[Integrate[E^(-x),{x,1,Infinity}]], N[Integrate[E^(-x),{x,1,10}]]}
```

$Out[36]=$
```
{0.367879, 0.367834}
```

Almost the same. For most practical purposes, and especially for government work, there is no difference between $\int_1^\infty e^{-x}dx$ and $\int_1^{10} e^{-x}dx$.

■ B.5) The integral of the sum is the sum of the integrals

Calculus&Mathematica is pleased to acknowledge the heavy influence of Emil Artin's book *Calculus with Analytic Geometry* (notes by G. B. Seligman) on this problem.

B.5.a)   How do you know that $\int_a^t (f[x] + g[x])\ dx = \int_a^t f[x]\ dx + \int_a^t g[x]\ dx$?

Answer: If you try to interpret this in terms of areas, this fact is not at all obvious; but the fundamental formula makes it a snap. Regard $f[x]$ as the derivative of another function $F[x]$, and regard $g[x]$ as the derivative of another function $G[x]$ so that $f[x] = F'[x]$ and $g[x] = G'[x]$. The fundamental formula says

$$\int_a^t f[x]\ dx = \int_a^t F'[x]\ dx = F[t] - F[a]$$

and

$$\int_a^t g[x]\ dx = \int_a^t G'[x]\ dx = G[t] - G[a].$$

Notice that $f[x] + g[x]$ is the derivative of $F[x] + G[x]$; so

$$\int_a^t f[x] + g[x]\ dx = \int_a^t F'[x] + G'[x]\ dx$$
$$= (F[t] + G[t]) - (F[a] + G[a])$$

$$= (F[t] - F[a]) + (G[t] - G[a])$$

$$= \int_a^t f[x]\, dx + \int_a^t g[x]\, dx.$$

There it is: $\int_a^t f[x] + g[x]\, dx = \int_a^t f[x]\, dx + \int_a^t g[x]\, dx$. The fundamental formula reduced this idea to a session in mindless symbol pushing. Another reason why the fundamental formula is so fundamental.

B.5.b) How do you know that $\int_a^t (f[x] - g[x])\, dx = \int_a^t f[x]\, dx - \int_a^t g[x]\, dx$?

Answer: More symbol pushing:

$$\int_a^t (f[x] - g[x])\, dx = \int_a^t (f[x] + (-g[x]))\, dx$$

$$= \int_a^t f[x]\, dx + \int_a^t (-g[x])\, dx$$

$$= \int_a^t f[x]\, dx - \int_a^t g[x]\, dx.$$

Nothing to get excited about, but now it's on the record.

■ B.6) Integrating backward: $\int_t^a g[x]\, dx = -\int_a^t g[x]\, dx$

B.6.a) In all cases you've seen so far, you've always looked at $\int_a^t g[x]\, dx$ where $a \le t$.

What do folks mean when they write $\int_a^t g[x]\, dx$ in the case that $t < a$?

Answer: This is a good question because if $t < a$, then the integral $\int_a^t g[x]\, dx$ does not have an obvious physical interpretation. Mathematicians and scientists worldwide have settled this via mutual convention by agreeing that if $t < a$, then $\int_a^t g[x]\, dx = -\int_t^a g[x]\, dx$. Check this out:

In[37]:=
```
Clear[g,x]; g[x_] = x^2; a = 1; t = 3;
{Integrate[g[x],{x,a,t}],Integrate[g[x],{x,t,a}]}
```

Out[37]=
$$\{\frac{26}{3},\ -(\frac{26}{3})\}$$

Yessiree Bob. $\int_a^t g[x]\, dx = -\int_t^a g[x]\, dx$. This agreement expedites calculation because it makes it true that $f[t] - f[a] = \int_a^t f'[x]\, dx$ regardless of whether $t \ge a$ or $t < a$. It's too bad that other worldwide issues cannot be dealt with in the same spirit of cooperation that worldwide scientific issues have received.

Tutorials

■ T.1) Getting the feel of the fundamental formula by using it to calculate integrals

Newton and Leibniz are called the founders of calculus because they cracked the code of the integral by writing down the fundamental formula of calculus

$$f[t] - f[a] = \int_a^t f'[x]\,dx$$

and, in one line of writing, changed the character of science. The fundamental formula is so important that *Mathematica* knows it cold:

In[1]:=
```
Clear[f,x,t,a,Derivative]
Integrate[f'[x],{x,a,t}]
```
Out[1]=
```
-f[a] + f[t]
```

The fundamental formula of calculus $f[t] - f[a] = \int_a^t f'[x]\,dx$ is your calculating jewel because it gives you a powerful, systematic way of calculating the exact value of integrals without resorting to approximations. And using it is a real pleasure. To attempt to calculate $\int_a^t g[x]\,dx$ for a given function $g[x]$, you find another function $f[x]$ such that $f'[x] = g[x]$ and then you just calculate

$$\int_a^t g[x]\,dx = \int_a^t f'[x]\,dx = f[t] - f[a].$$

In principle, nothing could be simpler, and you should not have to depend on the machine to do the simple cases.

To try it out, look at *Mathematica*'s calculation of $\int_3^9 x^2\,dx$:

In[2]:=
```
Clear[x]
Integrate[x^2,{x,3,9}]
```
Out[2]=
```
234
```

T.1.a.i)   Explain how *Mathematica* exploited the fundamental formula to do the calculation.

Answer: The fundamental formula says $f[t] - f[a] = \int_a^t f'[x]\,dx$. To use the fundamental formula to calculate $\int_3^9 x^2\,dx$, *Mathematica* looked for a function

$f[x]$ with $f'[x] = x^2$.

It found $f[x] = x^3/3$:

In[3]:=
```
Clear[x,f]; f[x_] = (x^3)/3; f'[x]
```
Out[3]=
```
 2
x
```

Good. Now using the fundamental formula, you get

$$f[9] - f[3] = \int_3^9 f'[x]\,dx = \int_3^9 x^2 dx,$$

Mathematica proclaims that $\int_3^9 x^2 dx$ is given by:

In[4]:=
```
f[9] - f[3]
```
Out[4]=
```
234
```

This is right on target:

In[5]:=
```
Integrate[x^2,{x,3,9}]
```
Out[5]=
```
234
```

Beautiful and simple.

In mathematics, most beautiful ideas are simple.

T.1.a.ii) | Use the fundamental formula to calculate $\int_0^{\pi/2} \cos[5\,x]\,dx$ and explain what the calculation measures.

Answer: The fundamental formula says $f[t] - f[a] = \int_a^t f'[x]\,dx$. To use the fundamental formula to calculate $\int_0^{\pi/2} \cos[5\,x]\,dx$, look for a function $f[x]$ with $f'[x] = \cos[5\,x]$. A moment's reflection tells you that $f[x] = \sin[5\,x]/5$ will do:

In[6]:=
```
Clear[x,f]; f[x_] = Sin[5 x]/5; f'[x]
```
Out[6]=
```
Cos[5 x]
```

Good. The fundamental formula,

$$f\left[\frac{\pi}{2}\right] - f[0] = \int_0^{\pi/2} f'[x]\,dx = \int_0^{\pi/2} \cos[x]\,dx,$$

tells you that $\int_0^{\pi/2} \cos[5\,x]\,dx$ is given by:

In[7]:=
```
f[Pi/2] - f[0]
```
Out[7]=
```
1
-
5
```

Check:

In[8]:=
```
Integrate[Cos[5 x],{x,0,Pi/2}]
```

Out[8]=

$$\frac{1}{5}$$

Nailed it. The calculation $\int_0^{\pi/2} \cos[5\,x]\,dx = 1/5$ means that the positive part of the shaded area below measures out to exactly one-fifth square unit more than the negative part:

In[9]:=
```
FilledPlot[Cos[5 x],{x,0,Pi/2},
AspectRatio->Automatic,
AxesLabel->{"x","y"},
PlotLabel->"y = Cos[5 x]"];
```

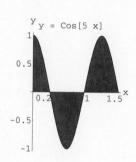

T.1.b) This problem appears only in the electronic version.

■ T.2) Velocity, acceleration, and the fundamental formula

T.2.a) A potted plant is thrown from a window of an apartment building, with an initial velocity of 2 feet per second directly down. The sill of the window is 215 feet above the sidewalk below.

> How long does it take for the plant to hit the sidewalk?
>
> With what velocity does it hit the sidewalk?

Answer: Acceleration due to gravity is 32 ft per second per second. Agree that the positive direction is up; this means (in feet and seconds):

In[10]:=
```
Clear[accel,vel,f,t,x]
accel[t_] = -32
```

Out[10]=
```
-32
```

Because $\text{vel}'[t] = \text{accel}[t]$, the fundamental formula guarantees that

$$\text{vel}[t] - \text{vel}[0] = \int_0^t \text{vel}'[x]\,dx = \int_0^t \text{accel}[x]\,dx.$$

This means $\text{vel}[t] = \text{vel}[0] + \int_0^t \text{accel}[x]\,dx$:

In[11]:=
```
vel[0] = -2; vel[t_] = vel[0] + Integrate[accel[x],{x,0,t}]
```
Out[11]=
```
-2 - 32 t
```

Put

$f[t]$ = plant's height above sidewalk t seconds after it was thrown.

Because $f'[t] = \mathrm{vel}[t]$ the fundamental formula guarantees that

$$f[t] - f[0] = \int_0^t f'[x]\, dx = \int_0^t \mathrm{vel}[x]\, dx.$$

This means $f[t] = f[0] + \int_0^t \mathrm{vel}[x]\, dx$:

In[12]:=
```
f[0] = 215; f[t_] = f[0] + Integrate[vel[x],{x,0,t}]
```
Out[12]=
```
            2
215 - 2 t - 16 t
```

To get the time it takes for the plant to hit the sidewalk, look at:

In[13]:=
```
N[Solve[f[t] == 0]]
```
Out[13]=
```
{{t -> 3.60375}, {t -> -3.72875}}
```

The potted plant takes 3.6 seconds to fall to the sidewalk. To get the velocity at impact, look at:

In[14]:=
```
vel[3.6]
```
Out[14]=
```
-117.2
```

It hits with a velocity of 117 feet per second, which is a pretty high speed. In miles per hour this translates to:

In[15]:=
```
vel[3.6] (60^2)/5280
```
Out[15]=
```
-79.9091
```

Almost 80 mph down! A good helmet would not be much protection against this.

T.2.b) Here's a line and a reference marker:

In[16]:=
```
a = -3; b = 3;
line = Graphics[{Thickness[0.02],
Line[{{0,a},{0,b}}]}];
marker = Graphics[{Red,Thickness[0.01],
Line[{{-0.25,0},{0.25,0}}]}];
Show[line,marker,
PlotRange->{{-2,2},{a,b}}];
```

An object moves on this line with the positive direction to the top. When you turn on the timer ($t = 0$), the object is 1.0 units below the marker. And t seconds after you turn on the timer, the object is moving with velocity

$$\text{vel}[t] = 3 \sin[4\,t] - 4 \cos[3\,t] \text{ units per second.}$$

Here a positive vel$[t]$ means the object is moving up and a negative vel$[t]$ means the object is moving down. Put

$$f[t] = \text{distance of the object above the marker } t \text{ seconds}$$
$$\text{after you turn on the timer.}$$

> Use the fundamental formula to help yourself come up with a formula for $f[t]$. Analyze the formula of $f[t]$ to describe the oscillatory nature of the motion. Make a little movie indicating one full oscillation of the object.

Answer: Here is what you know:

→ $f[0] = -1.0$ because that's where the object is when $t = 0$.

→ $f'[t] = \text{vel}[t] = 3 \sin[4\,t] - 4 \cos[3\,t]$ because velocity is the derivative of distance.

→ $f[t] - f[0] = \int_0^t f'[x]dx = \int_0^t \text{vel}[x]\,dx$ This is what the fundamental formula says because $f'[t] = \text{vel}[t]$.

Now you're in business because this tells you $f[t] = f[0] + \int_0^t \text{vel}[x]\,dx$:

In[17]:=
```
Clear[x,t,vel,f]
vel[t_] = 3 Sin[4 t] - 4 Cos[3 t];
f[0] = -1.0;
f[t_] = f[0] + Integrate[vel[x],{x,0,t}]
```

Out[17]=
$$-0.25 - \frac{3 \cos[4\,t]}{4} - \frac{4 \sin[3\,t]}{3}$$

Here comes a plot:

In[18]:=
```
Plot[f[t],{t,0,4 Pi},
AxesLabel->{"t","distance above\n marker"}];
```

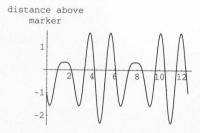

You can see the repeating pattern of the motion. To see why it works this way, look at the formula for $f[t]$, remembering that sin$[t]$ and cos$[t]$ repeat themselves every 2π, 4π, 6π, ... seconds.

In[19]:=
```
f[t]
```
Out[19]=

$$-0.25 - \frac{3 \, \text{Cos}[4 \, t]}{4} - \frac{4 \, \text{Sin}[3 \, t]}{3}$$

$\cos[4\,t]$ repeats itself every $\pi/2,\, 2\,(\pi/2),\, 3\,(\pi/2),\, \dots$ seconds.

In[20]:=
```
Clear[k]
Table[k Pi/2,{k,0,8}]
```
Out[20]=

$$\{0,\, \frac{\text{Pi}}{2},\, \text{Pi},\, \frac{3 \, \text{Pi}}{2},\, 2 \, \text{Pi},\, \frac{5 \, \text{Pi}}{2},\, 3 \, \text{Pi},\, \frac{7 \, \text{Pi}}{2},\, 4 \, \text{Pi}\}$$

$\sin[3\,t]$ repeats itself every $2\,\pi/3,\, 2\,(2\,\pi/3),\, 3\,(2\,\pi/3),\, \dots$ seconds.

In[21]:=
```
Clear[k]
Table[k 2 Pi/3,{k,0,8}]
```
Out[21]=

$$\{0,\, \frac{2 \, \text{Pi}}{3},\, \frac{4 \, \text{Pi}}{3},\, 2 \, \text{Pi},\, \frac{8 \, \text{Pi}}{3},\, \frac{10 \, \text{Pi}}{3},\, 4 \, \text{Pi},\, \frac{14 \, \text{Pi}}{3},\, \frac{16 \, \text{Pi}}{3}\}$$

The smallest common element in both lists is $2\,\pi$. This tells you that $f[t]$ completes one full oscillation every $2\,\pi$ seconds.

■ T.3) Some measurements based on the fundamental formula

T.3.a.i) Ten years ago ($t = 0$) you owed \$10,000 and you continued to borrow at a rate of $10000 \, e^{0.15t}$ additional dollars per year for $0 \le t \le 10$. If you put

> Debt$[t]$ = to the amount you owe t years from 10 years ago,

then you get Debt$'[t] = 10000 \, e^{0.15t}$.

> How much do you owe today?

Answer: The fundamental formula says Debt$[t]$ − Debt$[0] = \int_0^t$ Debt$'[x]\,dx$. So,

$$\text{Debt}[t] = \text{Debt}[0] + \int_0^t \text{Debt}'[x]\,dx.$$

In this problem:

In[22]:=
```
Clear[Debt,t,x]; Debt[0] = 10000;
Debt[t_] = Debt[0] + Integrate[10000 E^(0.15 x), {x,0,t}]
```

Out[22]=

$$-56666.7 + 66666.7\, E^{0.15\, t}$$

Here's a plot:

In[23]:=
```
Plot[Debt[t],{t,0,10},
 AxesLabel->{"t","debt"}];
```

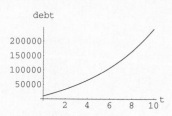

Ouch. Exponential growth is kicking in. Today you owe this many dollars:

In[24]:=
```
Debt[10]
```

Out[24]=
```
242113.
```

That's a lot of bread—almost a quarter of a million smackers. If you keep on going this way, here's how many dollars you'll owe 10 years from now:

In[25]:=
```
Debt[20]
```

Out[25]=

$$1.28237\ 10^{6}$$

Out of sight.

T.3.a.ii)   Get the formula for Debt[t] in part T.3.a.i) as the solution of a differential equation.

Answer:

In[26]:=
```
Clear[Debt,t,x,Derivative]
DSolve[{Debt'[t] == 10000 E^(0.15 t),Debt[0] == 10000},Debt[t],t]
```

Out[26]=

$$\{\{\text{Debt}[t]\ \text{->}\ -56666.7 + 66666.7\, E^{0.15\, t}\}\}$$

This is the same formula you got in part T.3.a.i). This is no accident because the fundamental formula says

$$\text{Debt}[t] - 10000 = \text{Debt}[t] - \text{Debt}[0]$$

$$= \int_0^t \text{Debt}'[x]\, dx$$

$$= \int_0^t 10000\, e^{0.15x}\, dx.$$

This is the same as saying $\text{Debt}'[t] = 10000\, e^{0.15t}$ and $\text{Debt}[0] = 10000$.

T.3.b.i) Ten years ago $(t = 0)$ you owed \$10,000 and you continued to borrow at a rate of $10000\,e^{0.25t}$ additional dollars per year for $0 \le t \le 10$. Fortunately your dowager auntie loved you enough to give you a trust fund that pays you at the rate of \$50,000 per year for life. Ten years ago you decided to use this trust money to help pay off your debt.

> Plot your debt as a function of t for $0 \le t \le 10$. Do you ever get out of debt? How much do you owe today? What are your prospects 10 years from now?

Answer: Measure t in years with $t = 0$ corresponding to 10 years ago. The fundamental formula says

$$\text{Debt}[t] - \text{Debt}[0] = \int_0^t \text{Debt}'[x]\,dx.$$

So,

$$\text{Debt}[t] = \text{Debt}[0] + \int_0^t \text{Debt}'[x]\,dx.$$

In this problem $\text{Debt}'[t] = 10000e^{0.25t} - 50000$:

In[27]:=
```
Clear[Debt,t,x];
Debt[0] = 10000;
Debt[t_] = Debt[0] + Integrate[10000 E^(0.25 x) - 50000,{x,0,t}]
```

Out[27]=
```
                0.25 t
  -30000. + 40000. E       - 50000 t
```

Here's a plot:

In[28]:=
```
Plot[Debt[t],{t,0,10},
 AxesLabel->{"t","debt"},
 PlotRange->All];
```

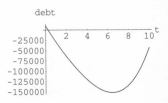

Your auntie put you into the chips. Today you owe nothing. Check out what lurks over the next 10 years: (Remember, $t = 10$ corresponds to today.)

In[29]:=
```
Plot[Debt[t],{t,10,20},
 AxesLabel->{"t","debt"},
 PlotRange->All];
```

Disaster. If you keep going on this way, here's how many dollars you'll owe 10 years from now:

In[30]:=
```
Debt[20]
```

Out[30]=

$$4.90653 \; 10^{6}$$

Almost five million dollars. Better find another rich aunt soon.

T.3.b.ii) | Get the formula for Debt[*t*] in part T.3.b.i) as the solution of a differential equation.

Answer: Ten years ago ($t = 0$) you owed \$10,000 and you continued to borrow at a rate of $10000 \, e^{0.25t}$ additional dollars per year for $0 \le t \le 10$. The trust fund pays you at the rate of \$50,000 per year for life.

In[31]:=
```
Clear[Debt,t,x,Derivative]
DSolve[{Debt'[t] == 10000 E^(0.25 t) - 50000,
Debt[0] == 10000},Debt[t],t]
```

Out[31]=

$$\{\{\text{Debt}[t] \to -30000. + 40000. \; E^{0.25 \; t} - 50000 \; t\}\}$$

This is the same formula you got in part T.3.b.i). This is no accident because the fundamental formula says

$$\text{Debt}[t] - 10000 = \text{Debt}[t] - \text{Debt}[0]$$

$$= \int_0^t \text{Debt}'[x] \, dx$$

$$= \int_0^t 10000 \, e^{0.25x} - 50000 \, dx.$$

This is the same as saying

$$\text{Debt}'[t] = 10000 e^{0.25t} - 50000$$

and

$$\text{Debt}[0] = 10000.$$

■ T.4) Area between curves

T.4.a) | Measure the area bounded by the curves $y = x^2$ and $y = x$.

Answer: To see what's going on, look at:

In[32]:=
```
Clear[f,g,x]; f[x_] = x^2;g[x_] = x;
Plot[{f[x],g[x]},{x,-1,2},
PlotStyle->{{Red,Thickness[0.01]},
{Blue,Thickness[0.01]}},
AxesLabel->{"x","y"},
PlotRange->{-1/2,2},
AspectRatio->Automatic];
```

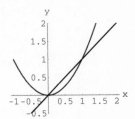

The goal is to measure the area of the crescent-shaped region to the right of the origin. See where the curves cross:

In[33]:=
```
Solve[f[x] == g[x],x]
```

Out[33]=
```
{{x -> 1}, {x -> 0}}
```

The area to be measured looks like this:

In[34]:=
```
a = 0; b = 1;
FilledPlot[{f[x],g[x]},{x,a,b},
AxesLabel->{"x","y"}];
```

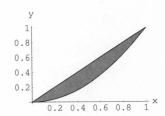

Here $g[x] = x$ is on top of $f[x] = x^2$. The measurement of the area of the crescent-shaped region in square units is

$$\int_0^1 g[x]\,dx - \int_0^1 f[x]\,dx = \int_0^1 (g[x] - f[x])\,dx :$$

In[35]:=
```
Integrate[(g[x] - f[x]),{x,a,b}]
```

Out[35]=
$$\frac{1}{6}$$

The fundamental formula at your service. The only trick is to be aware of where the curves cross, and be aware of which curve is on top and which curve is on the bottom.

T.4.b.i) Here's a plot of
$$f[x] = \frac{25.9\,e^x}{0.35 + 0.76\,e^x}$$
with a plot of
$$g[x] = 0.23\,e^x + 0.45\,e^{-x} - 12.20.$$

In[36]:=
```
Clear[f,g,x]
f[x_] = (25.9 E^x)/( 0.35 + 0.76 E^x);
g[x_] = 0.23 E^x + 0.45 E^(-x) - 12.20;
Plot[{f[x],g[x]},{x,-4,6},
PlotStyle->{Thickness[0.015],Blue},
AxesLabel->{"x",""}];
```

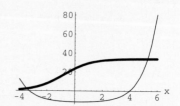

> Measure the area between the two curves.

Answer: First, get a better plot. The curves cross near $x = -3.5$ and again near $x = 5$. To get these crossovers accurately, you could try the solve instruction, but those exponentials will drive it up the wall. Instead use:

In[37]:=
```
a=x/.FindRoot[f[x] == g[x],{x,-3.5}][[1]]
```

Out[37]=
```
-3.4632
```

In[38]:=
```
b=x/.FindRoot[f[x] == g[x],{x,5}]
```

Out[38]=
```
5.30263
```

Here's a good picture of the area to be measured:

In[39]:=
```
goodplot =
FilledPlot[{f[x],g[x]},{x,a,b},
AxesLabel->{"x",""}];
```

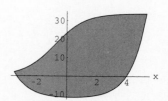

The $f[x]$ curve is on top; so the area measurement, in square units, is

$$\int_a^b f[x]\,dx - \int_a^b g[x]\,dx = \int_a^b (f[x] - g[x])\,dx :$$

In[40]:=
```
NIntegrate[f[x] - g[x],{x,a,b}]
```

Out[40]=
```
251.358
```

NIntegrate is a reasonable choice here because *Mathematica* will have a hard time using the fundamental formula. It is not easy to find a function whose derivative is $f[x] - g[x]$.

Something neat happened here. The $g[x]$ curve is running negative most of the way; so the $-\int_a^b g[x]\,dx$ term is actually measuring a positive area, just as it should.

T.4.b.ii) This problem appears only in the electronic version.

■ T.5) Approximate calculation of $\int_a^\infty f[x]\,dx$

For $s > 0$, calculating $\int_0^\infty e^{-x}\sin[x]\,dx$ is easy. You just look at the limiting case of $\int_0^t e^{-x}\sin[x]\,dx$ as $t \to \infty$:

In[41]:=
```
Clear[x,t]
Integrate[E^(-x) Sin[x],{x,0,t}]
```

Out[41]=

$$\frac{1}{2} - \frac{\text{Cos}[t]}{2\,E^t} - \frac{\text{Sin}[t]}{2\,E^t}$$

So,

$$\int_0^\infty e^{-x}\sin[x]\,dx = \lim_{t\to\infty}\int_0^t e^{-x}\sin[x]\,dx = \lim_{t\to\infty}\left(\frac{1}{2} - \frac{\cos[t]}{2\,e^t} - \frac{\sin[t]}{2\,e^t}\right)$$

$$= \frac{1}{2} + 0 + 0$$

$$= \frac{1}{2}$$

because the exponential e^t in the denominators of the second and third terms above dominates and sends these terms to 0 as $t \to \infty$.

T.5.a) Trying the same approach to calculate

$$\int_0^\infty \frac{e^{-x^2}}{\sqrt{1 + \sin[x]^2}}\,dx$$

doesn't work because nobody can obtain the exact value of $\int_0^t e^{-x^2}/\sqrt{1 + \sin[x]^2}\,dx$:

In[42]:=
```
Integrate[E^(-x^2)/Sqrt[1 + Sin[x]^2],{x,0,t}]
```

Out[42]=

$$\text{Integrate}\left[\frac{1}{E^{x^2}\,\text{Sqrt}[1 + \text{Sin}[x]^2]},\ \{x,\ 0,\ t\}\right]$$

In spite of this setback, come up with a reasonably accurate calculation of

$$\int_0^\infty \frac{e^{-x^2}}{\sqrt{1 + \sin[x]^2}}\,dx.$$

Answer: Because

$$\int_0^\infty \frac{e^{-x^2}}{\sqrt{1 + \sin[x]^2}}\, dx = \lim_{t \to \infty} \int_0^t \frac{e^{-x^2}}{\sqrt{1 + \sin[x]^2}}\, dx,$$

it's a good idea to examine what happens to $\int_0^t e^{-x^2}/\sqrt{1 + \sin[x]^2}\, dx$ for some large values of t:

In[43]:=
```
NIntegrate[E^(-x^2)/Sqrt[1 + Sin[x]^2],{x,0,5}]
```

Out[43]=
  0.786831

In[44]:=
```
NIntegrate[E^(-x^2)/Sqrt[1 + Sin[x]^2],{x,0,10}]
```

Out[44]=
  0.786831

In[45]:=
```
NIntegrate[E^(-x^2)/Sqrt[1 + Sin[x]^2],{x,0,100}]
```

Out[45]=
  0.786831

In[46]:=
```
NIntegrate[E^(-x^2)/Sqrt[1 + Sin[x]^2],{x,0,1000}]
```

Out[46]=
  0.786831

An educated, seat-of-the-pants estimate is that to about six decimals

$$\int_0^\infty \frac{e^{-x^2}}{\sqrt{1 + \sin[x]^2}}\, dx = \int_0^5 \frac{e^{-x^2}}{\sqrt{1 + \sin[x]^2}}\, dx = 0.786831.$$

Here is a way to get even more confidence in this educated guess: Note that

$$\int_0^\infty \frac{e^{-x^2}}{\sqrt{1 + \sin[x]^2}}\, dx = \int_0^5 \frac{e^{-x^2}}{\sqrt{1 + \sin[x]^2}}\, dx + \int_5^\infty \frac{e^{-x^2}}{\sqrt{1 + \sin[x]^2}}\, dx.$$

So

$$\int_0^5 e^{-x^2}/\sqrt{1 + \sin[x]^2}\, dx$$

estimates

$$\int_0^\infty e^{-x^2}/\sqrt{1 + \sin[x]^2}\, dx$$

within an error of no more than

$$\int_5^\infty e^{-x^2}/\sqrt{1 + \sin[x]^2}\, dx.$$

But the error estimate is

$$\int_5^\infty \frac{e^{-x^2}}{\sqrt{1+\sin[x]^2}}\, dx < \int_5^\infty e^{-x^2}\, dx,$$

and this last integral is easy to calculate:

In[47]:=
```
esterror = Integrate[E∧(-x∧2),{x,5,Infinity}]
```

Out[47]=
```
Sqrt[Pi] (1 - Erf[5])
─────────────────────
          2
```

Go to decimals:

In[48]:=
```
N[esterror,10]
```

Out[48]=
```
                    -12
  1.362517928 10
```

This tells you that

$$0 < \int_5^\infty \frac{e^{-x^2}}{\sqrt{1+\sin[x]^2}}\, dx < \int_5^\infty e^{-x^2}\, dx < 10^{-12}.$$

So,

$$\int_0^5 \frac{e^{-x^2}}{\sqrt{1+\sin[x]^2}}\, dx$$

estimates

$$\int_0^\infty \frac{e^{-x^2}}{\sqrt{1+\sin[x]^2}}\, dx$$

to at least 12 accurate decimals. This makes everyone really comfortable with the following estimate of

$$\int_0^\infty \frac{e^{-x^2}}{\sqrt{1+\sin[x]^2}}\, dx :$$

In[49]:=
```
NIntegrate[E∧(-x∧2)/Sqrt[1 + Sin[x]∧2],
{x,0,5},AccuracyGoal->8]
```

Out[49]=
```
  0.786831
```

T.5.b) This problem appears only in the electronic version.

■ T.6)   The fundamental formula and its relation to differential equations

T.6.a)   A certain function $f[x]$ has derivative
$$f'[x] = e^x - 2$$
and it is known that
$$f[2] = 7.$$

> Use the fundamental formula to find a formula for $f[x]$ and plot.
>
> Confirm your results by solving the differential equation.

Answer:

In[50]:=
```
Clear[f,fprime,x,t]; fprime[x_] = E^x - 2
```
Out[50]=
```
      x
-2 + E
```

The fundamental formula says that $f[x] - f[2] = \int_2^x f'[t]\,dt$. So,
$$f[x] = f[2] + \int_2^x f'[t]\,dt.$$
This gives you the formula for $f[x]$:

In[51]:=
```
f[x_] = (f[2] + Integrate[fprime[t],{t,2,x}])/.f[2]->7
```
Out[51]=
```
       2    x
11 - E  + E  - 2 x
```

Not very hard. Check:

In[52]:=
```
Clear[y]
DSolve[{y'[x] == fprime[x],y[2] == 7},y[x],x]
```
Out[52]=
```
                2    x
{{y[x] -> 11 - E  + E  - 2 x}}
```

Good. On the money. Here comes a plot:

In[53]:=
```
Plot[f[x],{x,-2,3},
PlotStyle->
{{Thickness[0.01],Red}},
AxesLabel->{"x","f[x]"}];
```

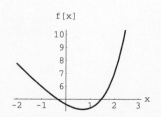

Look at that exponential term kick in.

T.6.b.i) Another function $g[x]$ has derivative

$$g'[x] = 2\,x - \frac{36}{x^3},$$

and it is known that

$$g[3] = -8.$$

Find a formula for $g[x]$ and plot.

Answer:

In[54]:=
```
Clear[gprime,x]
gprime[x_] = 2 x - 36/x^3
```
Out[54]=
$$\frac{-36}{x^3} + 2\,x$$

The fundamental formula says $g[x] - g[a] = \int_a^x g'[t]\,dt$. So, $g[x] = g[a] + \int_a^x g'[t]\,dt$. This gives you the formula for $g[x]$:

In[55]:=
```
Clear[g,t]; a = 3; g[a] = -8;
g[x_] = g[a] + Integrate[gprime[t],{t,a,x}]
```
Out[55]=
$$-19 + \frac{18}{x^2} + x^2$$

Check:

In[56]:=
```
Clear[y]
DSolve[{y'[x] == gprime[x],y[3] == -8},y[x],x]
```
Out[56]=
$$\{\{y[x] \to -19 + \frac{18}{x^2} + x^2\}\}$$

Good. Here is what seems to be the plot of $g[x]$ for $-3 < x < 8$:

In[57]:=
```
Plot[g[x],{x,-3,8},
PlotStyle->
{{Thickness[0.015],Blue}},
AxesLabel->{"x","g[x]"}];
```

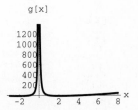

Look at that big blow-up at $x = 0$.

T.6.b.ii) Is the plot given in part T.6.b.i) entirely correct?

Answer: No way.

T.6.b.iii) What was wrong with the plot given in part T.6.b.i)? Give a correct plot.

Answer: You got the formula for $g[x]$ through the integral

$$g[x] = g[a] + \int_a^x g'[t]\, dt = g[3] + \int_3^x g'[t]\, dt.$$

Look at the formula for $g'[t]$:

In[58]:=
```
Clear[t]
gprime[t]
```

Out[58]=
```
-36
───  + 2 t
 3
t
```

The fact that $g'[t]$ blows up at $t = 0$:

In[59]:=
```
gprime[0]
Power::infy:
                     -3
    Infinite expression 0    encountered.
```

Out[59]=
```
ComplexInfinity
```

tells you that $\int_3^x g'[t]\, dt$ is meaningless for $x < 0$. Consequently, the formula:

In[60]:=
```
g[x]
```

Out[60]=
```
        18    2
-19 +  ──  + x
        2
       x
```

you got above is correct only for $x > 0$, and there is no way to determine a formula for $g[x]$ for $x \leq 0$. Here is a good representative correct plot of $g[x]$.

In[61]:=
```
Plot[g[x],{x,0.5,8},
 PlotStyle->
 {{Thickness[0.015],Blue}},
 AxesLabel->{"x","g[x]"}];
```

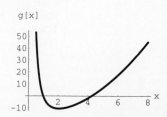

■ T.7) The indefinite integral $\int g[x]\,dx$

Given a function $g[x]$, you try to calculate the integral $\int_a^t g[x]\,dx$ by finding another function $f[x]$ such that $f'[x] = g[x]$ and then you calculate $\int_a^t g[x]\,dx = f[t] - f[a]$. In this set-up, most good old folks call the function $f[x]$ the antiderivative or the indefinite integral of $g[x]$. Over the years, it has become common to write $\int g[x]\,dx$ to signify a function whose derivative is $g[x]$. But this notation is loaded with danger, and it's best to avoid using it except on very special occasions.

T.7.a) For given constants a and b, one of the expressions

$$\int_a^b g[x]\,dx \qquad \text{and} \qquad \int g[x]\,dx$$

stands for a number; the other stands for a function.

> Which is which?

Answer: They are different species of animals. If a and b are given numbers, then $\int_a^b g[x]\,dx$ is a *number*. If a is a given number and t is allowed to vary, then $\int_a^t g[x]\,dx$ is a *function* of t. But $\int g[x]\,dx$ is always a *function* of x because $D[\int g[x]\,dx, x] = g[x]$.

T.7.b) > What is the hidden danger in the notation $\int g[x]\,dx$?

Answer: Look at an example: It is correct to say that

$$\int \cos[x]\,dx = \sin[x]$$

because $D[\sin[x], x] = \cos[x]$. It is equally correct to say that

$$\int \cos[x]\,dx = \sin[x] + 5$$

because $D[\sin[x] + 5, x] = \cos[x]$. In fact, it is correct to say that if C is any constant, then

$$\int \cos[x]\,dx = \sin[x] + C$$

because $D[\sin[x] + C, x] = \cos[x]$.

This is the rub: When you talk to a friend about $\int \cos[x]\,dx$, you and your friend may have different constants in mind, and this makes communication blurry. Matters are even worse when the function $g[x]$ has singularities: You could say correctly that

$$\int \left(\frac{1}{x}\right) dx = \log[x] + C;$$

implicit in this statement is the proviso that $x > 0$. You could also say correctly that

$$\int \left(\frac{1}{x}\right) dx = \log[|x|] + C$$

with the implicit proviso that $x \neq 0$. Many traditional calculus texts insist on

$$\int \left(\frac{1}{x}\right) dx = \log[|x|] + C,$$

but they are not exactly correct when they demand this. The trouble is that you could take two different constants and say correctly that

$$\int \left(\frac{1}{x}\right) dx = \log[x] + C_1 \qquad \text{for } x > 0$$

and

$$\int \left(\frac{1}{x}\right) dx = \log[-x] + C_2 \qquad \text{for } x < 0.$$

Old-style calculus books do not mention this. The inherent confusion about constants introduced by the notation $\int g[x]\,dx$ is the reason that computer mathematics processors sometimes screw up integrals. You will see this notation in Calculus&Mathematica only rarely.

T.7.c)   How does *Mathematica* handle the confusion about the constants in the notation $\int g[x]\,dx$?

Answer: The *Mathematica* instruction Integrate$[g[x], x]$ tells *Mathematica* to try to find a function $f[x]$ such that $f'[x] = g[x]$; that is, $f[x] = \int g[x]\,dx$.

Let's try it: For $\int x^2\,dx$, use:

In[62]:=
    Integrate[x∧2,x]

Out[62]=
$$\frac{x^3}{3}$$

For $\int e^x\,dx$, use:

In[63]:=
```
Integrate[E^x,x]
```

Out[63]=
```
 x
E
```

For $\int \sin[x]\,dx$, use:

In[64]:=
```
Integrate[Sin[x],x]
```

Out[64]=
```
-Cos[x]
```

For $\int 1/\left(1+x^2\right)\,dx$, use:

In[65]:=
```
Integrate[1/(1 + x^2),x]
```

Out[65]=
```
ArcTan[x]
```

For $\int \sqrt{1-x^2}\,dx$, use:

In[66]:=
```
Integrate[1/Sqrt[1 - x^2],x]
```

Out[66]=
```
ArcSin[x]
```

Evidently, *Mathematica* deals with the problem of the constants by leaving the constants off altogether. Certainly not a bad compromise. It's fun to realize that this convention is a total insult to some moralistic, out-of-date math teachers, who get their jollies by deducting points when the student doesn't include the constant.

Give It a Try

Experience with the starred ($\star$) problems will be especially beneficial for understanding later lessons.

■ G.1) Calculating integrals by solving differential equations*

G.1.a) Here are some experiments that show formulas for

$$\int_a^t g[x]\,dx$$

side-by-side with the solution of the differential equation

$$y'[t] = g[t] \qquad \text{with } y[a] = 0$$

for various choices of $g[x]$ and a:

Experiment 1:

In[1]:=
```
Clear[x,t,y,g,Derivative]
a = 0; g[x_] = x^2;
{Integrate[g[x],{x,a,t}],DSolve[{y'[t] == g[t],y[a] == 0},y[t],t]}
```

Out[1]=
$$\{\frac{t^3}{3}, \{\{y[t] \rightarrow \frac{t^3}{3}\}\}\}$$

Experiment 2:

In[2]:=
```
Clear[x,t,y,g,Derivative]
a = 1; g[x_] = x^2;
Together[{Integrate[g[x],{x,a,t}],DSolve[{y'[t] == g[t],y[a] == 0},y[t],t]}]
```

Out[2]=
$$\{\frac{-1 + t^3}{3}, \{\{y[t] \rightarrow \frac{-1 + t^3}{3}\}\}\}$$

Experiment 3:

In[3]:=
```
Clear[x,t,y,g,Derivative]
a = Pi; g[x_] = Sin[x]^2;
{Integrate[g[x],{x,a,t}],DSolve[{y'[t] == g[t],y[a] == 0},y[t],t]}
```

Out[3]=
$$\{\frac{-Pi}{2} + \frac{t}{2} - \frac{Sin[2\ t]}{4}, \{\{y[t] \rightarrow \frac{-Pi}{2} + \frac{t}{2} - \frac{Sin[2\ t]}{4}\}\}\}$$

Experiment 4:

In[4]:=
```
Clear[x,t,y,g,Derivative]
a = 0; g[x_] = (2/Sqrt[Pi]) E^(-x^2);
{Integrate[g[x],{x,a,t}],DSolve[{y'[t] == g[t],y[a] == 0},y[t],t]}
```

Out[4]=
```
{Erf[t], {{y[t] -> Erf[t]}}}
```

Try some more experiments with new choices of a and $g[x]$. Explain why the results came out the way they did and why they will turn out this way no matter what function $g[x]$ you type in (unless *Mathematica* cannot do the integral or *Mathematica* screws up).

G.1.b)   Use the fundamental formula to explain why calculating the formula for the integral $\int_a^t g[x]\,dx$ is the same as the formula for the solution of the differential equation $y'[t] = g[t]$ with $y[a] = 0$.

G.1.c)   You need a plot of $\int_0^t e^{-x} \sin[x^2]\, dx$ for $1 \le t \le 3$. You try:

In[5]:=
```
Clear[x,g,t]; g[x_] = E^(-x) Sin[x^2];
Integrate[g[x],{x,0,t}]
```

Out[5]=

$$\text{Integrate}[\frac{\text{Sin}[x^2]}{E^x},\ \{x,\ 0,\ t\}]$$

This is not a deficiency of *Mathematica*. The problem here is that there is no possible clean formula for a function $f[x]$ whose derivative is $g[x]$.

Now look at these:

In[6]:=
```
Clear[f,x,t]
f[t_] := NIntegrate[g[x],{x,0,t}];
Plot[f[t],{t,1,3},
PlotStyle->{{PrussianBlue,
Thickness[0.015]}}];
```

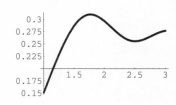

and

In[7]:=
```
Clear[y,t,Derivative]
solution = NDSolve[{y'[t] == g[t],
y[0] == 0},y[t],{t,1,3}];
y[t_] = y[t]/.solution[[1]];
Plot[y[t],{t,1,3},
PlotStyle->{{PrussianBlue,
Thickness[0.015]}}];
```

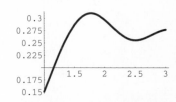

Explain why you can choose either one.

Which one ran faster? Which one do you prefer?

G.1.d.i)   In spite of its innocence, $x \tan[x]$ is not the derivative of any function $f[x]$ that can be expressed in terms of a clean formula.

How does this fact explain why the following instruction fails?

In[8]:=
```
Clear[x,t]; Integrate[x Tan[x],{x,-1,t}]
```

Out[8]=
```
Integrate[x Tan[x], {x, -1, t}]
```

G.1.d.ii)

> Use NDSolve to estimate the value of $\int_{-1}^{1} x \tan[x]\, dx$. Check yourself with NIntegrate.

G.1.e)

> Give a representative plot of the function $f[x]$ with $f'[x] = x + \sin[x]$ and $f[0] = 4$.

G.1.f)

> Use the fundamental formula to find a formula for $y[x]$ in terms of x given that $y''[x] = e^{-x}\cos[x]$ and given that the graph of $y = y[x]$ passes through the point $\{1.3, -1.57\}$ with an instantaneous growth rate (slope) of 0.9. Give a representative plot of y for $x \geq 0$.

■ G.2) How does *Mathematica* calculate an integral? How do you calculate an integral by hand?*

When you feed a formula for a function $g[x]$ into *Mathematica* and ask *Mathematica* to calculate $\int_a^t g[x]\, dx$, *Mathematica* looks for a function $f[t]$ with $f'[t] = g[t]$ and then it reports $\int_a^t g[x]\, dx = f[t] - f[a]$. Examples:

In[9]:=
```
Clear[g,x,t,a]; g[x_] = Cos[x];
Integrate[g[x],{x,a,t}]
```

Out[9]=
```
-Sin[a] + Sin[t]
```

In[10]:=
```
Clear[g,x,t,a]; g[x_] = x;
Integrate[g[x],{x,a,t}]
```

Out[10]=
$$-\frac{a^2}{2} + \frac{t^2}{2}$$

In[11]:=
```
Clear[g,x,t,a]; g[x_] = E^x;
Integrate[g[x],{x,a,t}]
```

Out[11]=
$$-E^a + E^t$$

In[12]:=
```
Clear[g,x,t,a];
g[x_] = (2/Sqrt[Pi])E^(-x^2);
Integrate[g[x],{x,a,t}]
```

Out[12]=
 -Erf[a] + Erf[t]

Once a function $f[t]$ with $f'[t] = g[t]$ is found, then the reported answer is guaranteed to be correct because the fundamental formula says that

$$\int_a^t g[x]\,dx = \int_a^t f'[x]\,dx = f[t] - f[a].$$

When you calculate integrals, you do the same thing. The only difference is that *Mathematica* has a great advantage over you because it is faster than you and has a bigger collection of functions to choose from than you have. Nevertheless, for simple cases, you should be able to beat *Mathematica* by doing integrals in your head faster than you can type them into *Mathematica*. Scientists who cannot calculate an integral like $\int_0^1 x^2\,dx$ in their heads quickly without machine help are labeled illiterate just as people who can't spell "cat" are labeled illiterate.

Find the exact values of the following integrals in your head; check yourself with *Mathematica*. Don't be afraid to take out some scratch paper.

G.2.a) Basic e^x integrals

G.2.a.i)
$$\int_a^t e^x\,dx$$

G.2.a.ii)
$$\int_a^t e^{2x}\,dx$$

G.2.a.iii)
$$\int_a^t e^{-3x}\,dx$$

G.2.a.iv)
$$\int_a^t e^{-x^2}\,2\,x\,dx$$

G.2.b) Basic $\cos[x]$ integrals

G.2.b.i)
$$\int_0^t \cos[x]\,dx$$

G.2.b.ii)
$$\int_0^t \cos[2\,x]\,dx$$

G.2.b.iii)
$$\int_0^t \cos[3\,x]\,dx$$

G.2.b.iv)
$$\int_a^t \cos[x^2]\,2\,x\,dx$$

G.2.c) Basic $\sin[x]$ integrals

G.2.c.i)
$$\int_0^t \sin[x]\,dx$$

G.2.c.ii)
$$\int_0^t \sin[2\,x]\,dx$$

G.2.c.iii)
$$\int_0^t \sin[3\,x]\,dx$$

G.2.c.iv)
$$\int_a^t \sin[x^2]\,2\,x\,dx$$

G.2.d) Mixing them up

G.2.d.i)
$$\int_0^{\pi/2} e^{\cos[x]}\,\sin[x]\,dx$$

G.2.d.ii)
$$\int_0^{\pi/2} e^{2\,\cos[x]}\,\sin[x]\,dx$$

G.2.d.iii)
$$\int_a^t \cos[e^x]\,e^x\,dx$$

G.2.e) Basic x^k integrals

G.2.e.i)
$$\int_a^t 1\,dx$$

G.2.e.ii)
$$\int_a^t x\,dx$$

G.2.e.iii)
$$\int_a^t x^2\,dx$$

G.2.e.iv)
$$\int_a^t x^3\,dx$$

G.2.e.v)
$$\int_a^t \left(x^3 - 4\,x^2 + 5\,x + 3\right)\,dx$$

G.2.e.vi)
$$\int_a^t x^{-2}\,dx$$

G.2.e.vii)
$$\int_a^t \left(2\,x^2 + 4\,x^{-2}\right)\,dx$$

G.2.e.viii)
$$\int_0^3 dx$$

G.2.f) Basic $\log[x]$ integrals

G.2.f.i)
$$\int_1^t \frac{1}{x}\,dx \qquad (\text{with } t > 0)$$

G.2.f.ii)
$$\int_a^t \frac{f'[x]}{f[x]}\,dx$$

G.2.f.iii)
$$\int_a^t \frac{4\,x^3}{x^4 + 1}\,dx$$

G.2.f.iv)
$$\int_a^t \frac{e^x}{2 + e^x}\,dx$$

G.2.f.v)

$$\int_{\pi/6}^{\pi/3} \frac{\cos[x]}{\sin[x]}\, dx$$

G.2.g) A ringer

$$\int_{-4}^{4} \sin[x^3]\, dx$$

■ G.3) Velocity and acceleration⋆

G.3.a)

A rock is dropped into a dry well and hits the bottom in 3.7 seconds. How deep is the well?

G.3.b.i) Here is a line and a reference marker:

In[13]:=
```
a =-6; b = 2;
line = Graphics[{Thickness[0.02],
Line[{{0,a},{0,b}}]}];
marker = Graphics[{Red,Thickness[0.01],
Line[{{-0.25,0},{0.25,0}}]}];
Show[line,marker,PlotRange->{{-2,2},{a,b}}];
```

An object moves on this line with the positive direction to the top. When you turn on the timer ($t = 0$), the object is 1.5 units below the marker. And t seconds after you turn on the timer, the object is moving with velocity

$$\text{vel}[t] = 3\cos[4\,t] - 4\sin[3\,t] \text{ units per second.}$$

Here a positive vel[t] means the object is moving up and a negative vel[t] means the object is moving down. Put

$$f[t] = \text{distance of the object above the marker } t \text{ seconds after}$$
$$\text{you turn on the timer.}$$

Use the fundamental formula to help come up with a formula for $f[t]$.

Analyze the formula of $f[t]$ to describe the oscillatory nature of the motion.

Make a little movie indicating a full oscillation of the object.

G.3.b.ii) In part G.3.b.i), the velocity of the object at time t is given by

$$\text{vel}[t] = 3\,\cos[4\,t] - 4\,\sin[3\,t] \text{ units/sec.}$$

The speed of the object is given by

$$\text{speed}[t] = |3\,\cos[4\,t] - 4\,\sin[3\,t]| \text{ units/sec.}$$

So, $\text{speed}[t]^2 = \text{vel}[t]^2$ and $\text{speed}[t]$ is greatest when $\text{vel}[t]^2$ is as big as possible, and $\text{speed}[t]$ is least when $\text{vel}[t]^2$ is as small as possible.

> Report on the greatest and smallest speed at which this object moves.

G.3.c) Here is another line and another reference marker:

```
In[14]:=
  line = Graphics[{Thickness[0.02],
  Line[{{-0.5,0},{1,0}}]}];
  marker = Graphics[{Red,Thickness[0.01],
  Line[{{0,0.2},{0,-0.2}}]}];
  Show[line,marker,PlotRange->{-0.3,0.3},
  AspectRatio->1/7];
```

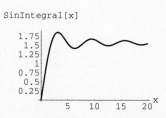

An object moves on this line. When you turn on the timer ($t = 0$), the object is 0.5 units to the right of the marker, moving with velocity 0.35 units per second to the left. And t seconds after you turn on the timer, the object's acceleration directed toward the right is given by $\text{accel}[t] = 2\sin[5\,t]$ units per second per second. Put $\text{vel}[t]$ = the velocity of the object to the right t seconds after you turn on the timer and $f[t]$ = the distance of the object from the right of the marker t seconds after you turn on the timer.

> Use the fundamental formula to give formulas for $\text{vel}[t]$ and $f[t]$. Then plot $\text{vel}[t]$ and $f[t]$, $\text{vel}[t]$ and $\text{accel}[t]$ for $0 \le t \le 12$ individually and show them together.
>
> Is this motion oscillatory? Make a little movie portraying the motion of the object.

■ G.4) Functions defined by integrals⋆

Here's a plot of a function you probably have never had the pleasure of meeting:

```
In[15]:=
  Clear[x]
  Plot[SinIntegral[x],{x,0,20},
  PlotStyle->{{Blue,Thickness[0.01]}},
  PlotRange->All,AxesLabel->
  {"x","SinIntegral[x]"}];
```

To learn the formula for SinIntegral[x], put $f[x] = $ SinIntegral[x] and look at $f'[x]$:

A bug in *Mathematica* 2.1 will not allow you to calculate $f'[x]$ with the usual prime notation.

In[16]:=
```
Clear[f,x]; f[x_] = SinIntegral[x]; D[f[x],x]
```

Out[16]=

$$\frac{\text{Sin[x]}}{\text{x}}$$

Now look at $f[0]$:

In[17]:=
```
f[0]
```

Out[17]=
```
0
```

The fundamental formula says $f[x] = f[0] + \int_0^x f'[t]\,dt$; so

$$\text{SinIntegral}[x] = 0 + \int_0^x \frac{\sin[t]}{t}\,dt = \int_0^x \frac{\sin[t]}{t}\,dt.$$

There you have the formula for SinIntegral[x]. This formula is as clean as it can be because the integral $\int_0^x \sin[t]/t\,dt$ can't be calculated in terms of other familiar functions.

G.4.a.i) Here's another new friend and its derivative:

In[18]:=
```
Clear[f,x]; f[x_] = Erf[x]; D[f[x],x]
```

Out[18]=

$$\frac{2}{\text{E}^{\text{x}^2}\ \text{Sqrt[Pi]}}$$

Here is $f[0]$:

In[19]:=
```
f[0]
```

Out[19]=
```
0
```

> Use the fundamental formula $f[x] = f[a] + \int_a^x f'[t]\,dt$ to write down a clean formula for $f[x] = \text{Erf}[x]$ in terms of an integral.

Just for your information: The shape of the plot of the derivative of Erf[x] is very famous. Take a look:

```
In[20]:=
Clear[f,x]
f[x_] = Erf[x]; Plot[f'[x],{x,-2,2},
PlotStyle->{{SteelBlue,Thickness[0.01]}},
AspectRatio->Automatic];
```

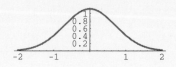

Stick around for a couple more lessons and you'll learn what you can do with this bell.

G.4.a.ii) Here's another new friend and its derivative:

```
In[21]:=
Clear[f,x]; f[x_] = Erfc[x]; D[f[x],x]
```

```
Out[21]=
        -2
   ───────────
      2
     x
    E   Sqrt[Pi]
```

Here is $f[0]$:

```
In[22]:=
f[0]
```

```
Out[22]=
1
```

> Use the fundamental formula $f[x] = f[a] + \int_a^x f'[t]\,dt$ to write down a clean formula for $f[x] = \mathrm{Erfc}[x]$ in terms of an integral.

G.4.a.iii) Here's yet another new friend and its derivative:

```
In[23]:=
Clear[f,x,k]; f[x_] = EllipticE[x,k]; D[f[x],x]
```

```
Out[23]=
                    2
Sqrt[1 - k Sin[x] ]
```

Here is $f[0]$:

```
In[24]:=
f[0]
```

```
Out[24]=
0
```

> Use the fundamental formula $f[x] = f[a] + \int_a^x f'[t]\,dt$ to write down a clean formula for $f[x] = \mathrm{EllipticE}[x,k]$ in terms of an integral.

G.4.a.iv) The suspicious function

$$f[x] = x^2 - 2\,x\log[1 + e^x] - 2\,\mathrm{PolyLog}[2, -e^x]$$

is so obscure that it doesn't even have its own name. Besides, who even knows what $\text{PolyLog}[2, -e^x]$ is? Nevertheless, *Mathematica* will give you a clean formula for $f'[x]$:

In[25]:=
```
Clear[f,x]
f[x_] = x^2 - 2 x Log[1 + E^x] - 2 PolyLog[2,-E^x];
Together[D[f[x],x]]
```

Out[25]=

$$\frac{2\ x}{1 + E^x}$$

Here is $f[0]$:

In[26]:=
```
f[0]
```

Out[26]=

$$\frac{\text{Pi}^2}{6}$$

> Use the fundamental formula $f[x] = f[a] + \int_a^x f'[t]\,dt$ to write down a clean formula for
> $$f[x] = x^2 - 2\,x\log[1 + e^x] - 2\,\text{PolyLog}[2, -e^x]$$
> in terms of an integral.

G.4.a.v) Here's a trusted old friend and its derivative:

In[27]:=
```
Clear[f,x]; f[x_] = Log[x]; D[f[x],x]
```

Out[27]=

$$\frac{1}{x}$$

Here is $f[1]$:

In[28]:=
```
f[1]
```

Out[28]=
```
0
```

> Use the fundamental formula $f[x] = f[a] + \int_a^x f'[t]\,dt$ to write down a clean formula for $f[x] = \log[x]$ in terms of an integral.

G.4.a.vi) Here's an old friend and its derivative:

In[29]:=
```
Clear[f,x]; f[x_] = ArcTan[x]; D[f[x],x]
```
Out[29]=
$$\frac{1}{1 + x^2}$$

Here is $f[0]$:

In[30]:=
```
f[0]
```
Out[30]=
```
0
```

Use the fundamental formula $f[x] = f[a] + \int_a^x f'[t]\,dt$ to write down a clean formula for $f[x] = \text{Arctan}[x]$ in terms of an integral.

G.4.b.i) The fact that if $f[t] = \int_a^t g[x]\,dx$, then $f'[t] = g[t]$ is a welcome calculational fringe benefit because it allows you to find the derivative without calculating the integral! To get the hang of this, run the following cells:

In[31]:=
```
Clear[f,g,x,t,a]
g[x_] = Cos[Pi x];
f[t_] = Integrate[g[x],{x,a,t}];
{f'[t],g[t]}
```
Out[31]=
```
{Cos[Pi t], Cos[Pi t]}
```

In[32]:=
```
Clear[f,g,x,t,a]
g[x_] = E^(2 x);
f[t_] = Integrate[g[x],{x,a,t}];
{f'[t],g[t]}
```
Out[32]=
$$\{E^{2t}, E^{2t}\}$$

In[33]:=
```
Clear[f,g,x,t,a]
g[x_] = x^4 - 9;
f[t_] = Integrate[g[x],{x,a,t}];
{f'[t],g[t]}
```
Out[33]=
$$\{-9 + t^4, -9 + t^4\}$$

Explain what's going on.

G.4.b.ii) Put

$$f[t] = \int_0^t \sin[x^4]\,dx$$

and give a formula for $f'[t]$.

G.4.b.iii) Put

$$f[t] = \int_0^{t^2} \sin[x^4]\,dx$$

and give a formula for $f'[t]$.

G.4.c) Put

$$f[t] = \int_{-1}^t (x+1)^2 \tan[x]\,dx$$

and give a formula for $f'[t]$.

Find the t that minimizes $f[t]$ on the interval $[-1,1]$.

G.4.d.i) Find a solution $f[x]$ of

$$e^{2t} - \int_0^t f[x]\,dx = 1$$

that works for all t's.

G.4.d.ii) Explain why there is no solution $f[x]$ of

$$\sin[t] - \int_0^t f[x]\,dx = 1$$

that works for all t's.

■ G.5) Plotting $(f[t+h] - f[t])/h$ and $g[t]$ when $f[t] = \int_a^t g[x]\,dx^\star$

This problem appears only in the electronic version.

■ G.6) Some measurements coming from the fundamental formula[*]

G.6.a) The village water tank has a capacity of 625,000 cubic feet of water. Some vandals broke into the tank and dropped three dead pigs into the water, thus contaminating the village water supply. The village officials spring into action by emptying and cleaning the tank. They pump new fresh water into the tank at a rate of

$$\frac{210000\,e^t}{4.1 + 1.2\,e^t} \text{ cubic feet per day}$$

t days after the refilling process begins. Agree that

Water$[t]$ = cubic feet of water in the tank t days from
the time the refilling process begins.

> Assuming no water is pumped out of the tank, give a plot Water$[t]$.
>
> About how many days will it take to fill the tank?

G.6.b.i) The village water tank has a capacity of 625,000 cubic feet of water. This tank supplies water to 4350 households, which use an average of 25 cubic feet of water per day. All of a sudden, the village officials realize that the pump that fills the tank has failed. At this time the tank contains 157,000 cubic feet of water.

> How long do they have to get the pump repaired before the tank runs dry?

G.6.b.ii) The village officials succeed in repairing the pump before the tank runs dry. When they turn on the pump, the tank contains only 61,000 cubic feet of water. The newly repaired pump can throw new water into the tank at a rate of

$$\frac{210000\,e^t}{4.1 + 1.2\,e^t} \text{ cubic feet per day}$$

where t is measured in days since the officials turned on the newly repaired pump.

> Is this pump fast enough to keep the tank from running dry while the tank continues to supply the 4350 households that use an average of 25 cubic feet of water per day?

G.6.b.iii)
> How many extra cubic feet of water over and above the water pumped in do the village officials need to add to the tank to keep the tank from running dry?

■ G.7) $\int_a^\infty f[x]\,dx$: Exact and approximate calculations⋆

G.7.a)   Here is *Mathematica*'s calculation of $\int_0^t e^{-x}\cos[3\,x]\,dx$:

```
In[34]:=
  Clear[x,t]
  Integrate[E^(-x) Cos[3 x],{x,0,t}]
```

```
Out[34]=
```
$$\frac{1}{10} - \frac{\text{Cos}[3\ t]}{10\ \text{E}^t} + \frac{3\ \text{Sin}[3\ t]}{10\ \text{E}^t}$$

> Study the limiting behavior of $\int_0^t e^{-x}\cos[3\,x]\,dx$ as $t \to \infty$ to calculate the exact value of $\int_0^\infty e^{-x}\cos[3\,x]\,dx$.

G.7.b)   Here is *Mathematica*'s calculation of

$$\int_e^t \frac{1}{x\,\log[x]^2}\,dx :$$

```
In[35]:=
  Clear[x,t]
  Integrate[1/(x Log[x]^2),{x,E,t}]
```

```
Out[35]=
```
$$1 - \frac{1}{\text{Log}[t]}$$

> Calculate the exact value of $\int_e^\infty 1/\left(x\,\log[x]^2\right)\,dx$ by examining the behavior of $\int_e^t 1/\left(x\,\log[x]^2\right)\,dx$ as $t \to \infty$.

G.7.c)   Here is *Mathematica*'s calculation of

$$\int_1^\infty \left(\frac{\sin[x/5]}{x}\right)\,dx :$$

```
In[36]:=
  Clear[x]
  integral = Integrate[Sin[x/5]/x, {x,1,Infinity}]
```

```
Out[36]=
```
$$\frac{\text{Pi} - 2\ \text{SinIntegral}[\frac{1}{5}]}{2}$$

Here is

$$\int_1^\infty \left(\frac{\sin[x/5]}{x}\right) dx$$

to 12 accurate decimals:

In[37]:=
```
twelveaccurate = N[integral,13]
```

Out[37]=
```
1.371240238269
```

How big do you have to make t to force

$$\int_1^t \left(\frac{\sin[x/5]}{x}\right) dx$$

to agree with

$$\int_1^\infty \left(\frac{\sin[x/5]}{x}\right) dx$$

to six accurate decimals?

G.7.d) Here is *Mathematica*'s calculation of $\int_0^\infty \sin[x^2]\, dx$:

In[38]:=
```
Clear[x]
integral = Integrate[Sin[x^2], {x,0,Infinity}]
```

Out[38]=
```
      Pi
Sqrt[--]
      8
```

Here is $\int_0^\infty \sin[x^2]\, dx$ to 10 accurate decimals:

In[39]:=
```
tenaccurate = N[integral,10]
```

Out[39]=
```
0.6266570687
```

How big do you have to make t to force $\int_0^t \sin[x^2]\, dx$ to agree with $\int_0^\infty \sin[x^2]\, dx$ to five accurate decimals?

G.7.e) Come up with a reasonably accurate estimate of

$$\int_1^\infty \frac{e^{-x}}{\sqrt{1+x^4}}\, dx.$$

Discuss the accuracy of your estimate.

■ G.8) Waterloo Tiles

Calculus&*Mathematica* is pleased to acknowledge that this problem was adapted from the *Waterloo Maple Calculus Workbook*, University of Waterloo, 1988.

The Waterloo Tile Co. is designing 1 foot by 1 foot ceramic tiles. On each tile, two cubic curves are running from the lower left-hand corner to the upper right-hand corner. They are positioned so that they trisect the square's right angles at the lower left-hand and the upper right-hand corners.

G.8.a.i)   Find the equations of the two cubic curves and plot them for $0 \le x \le 1$.

G.8.a.ii)   Between the cubic curves, a shade of red is to be applied to each tile. A shade of beige is to be applied to the rest of each tile.

Determine the relative amount of red and beige needed by calculating the ratio of the red area to the beige area.

G.8.b)   The Waterloo Tile Co. gets a different order for 1 foot by 1 foot ceramic tiles. On each tile, two fourth-degree polynomial curves are running from the lower left-hand corner to the upper right-hand corner, with one running above the other. They are positioned so that they trisect the square's right angles at the lower left-hand and the upper right-hand corners. The two curves cut the tile into three pieces of equal area.

Find the equations of the two curves and plot them for $0 \le x \le 1$.

■ G.9) Cases in which the fundamental formula fails and *Mathematica* screws up[★]

Students who have been programmed to calculate integrals by rote memorized procedures often run into the same problems that machines run into. They get the wrong answer because of misuse of the fundamental formula. So do mathematics processors like *Mathematica*. Look at this attempt at a calculation of

$$\int_{-1}^{3} \frac{1}{x^4}\, dx :$$

In[40]:=
```
Clear[x]
Integrate[1/x^4,{x,-1,3}]
```

Out[40]=

$$-\left(\frac{28}{81}\right)$$

Mathematica found a function $f[x] = -1/\left(3\,x^3\right)$ with $f'[x] = 1/x^4$.

In[41]:=
```
Clear[f,x]; f[x_] = -1/(3 x^3); f'[x]
```

Out[41]=

$$\frac{-4}{x}$$

Then *Mathematica* used the fundamental formula to report that

$$\int_{-1}^{3} \frac{1}{x^4}\,dx = \int_{-1}^{3} f'[x]\,dx = f[3] - f[-1] :$$

In[42]:=
```
f[3] - f[-1]
```

Out[42]=

$$-\left(\frac{28}{81}\right)$$

This looks fine and dandy, but it is dead wrong. Here are three reasons:

→ Surface level: $f'[x] = 1/x^4$ is always positive; so $\int_{-1}^{3} 1/x^4\,dx$ cannot be negative.

→ Surface level: NIntegrate refuses to estimate this integral:

In[43]:=
```
NIntegrate[f'[x],{x,-1,3}]
```
```
Power::infy:
                        -4
    Infinite expression 0    encountered.

Power::infy:
                        -4
    Infinite expression 0    encountered.

NIntegrate::inum:
    Integrand ComplexInfinity
        is not numerical at {x} = {0}.
```

Out[43]=
```
NIntegrate[f'[x], {x, -1, 3}]
```

→ Deeper level: $f'[x] = 1/x^4$ blows up at $x = 0$. It makes no sense to try to calculate $\int_{-1}^{3} f'[x]\,dx$ because $f'[x]$ has that big fat singularity at 0.

G.9.a.i) Look at *Mathematica*'s calculation of

$$\int_{-2}^{3} \frac{1}{x^2}\,dx :$$

In[44]:=
```
Clear[x]; Integrate[1/x^2,{x,-2,3}]
```

Out[44]=

$$-\left(\frac{5}{6}\right)$$

> Do you agree or disagree, and why?

G.9.a.ii) Look at *Mathematica*'s calculation of

$$\int_a^t \frac{1}{x^2}\,dx :$$

In[45]:=
```
Clear[x,a,t]; Integrate[1/x^2,{x,a,t}]
```

Out[45]=

$$\frac{1}{a} - \frac{1}{t}$$

> Is this right in the case that $0 < a \le t$?
>
> Is this right in the case that $a \le t < 0$?
>
> Is this right in the case that $a < 0 < t$?

G.9.b) Come up with your own example of a function $g[x]$ such that if you attempt to use the fundamental formula to calculate $\int_1^5 g[x]\,dx$, then you'll get garbage.

LESSON 2.03

Measurements

Basics

■ B.1) Measurements based on slicing and accumulating: Area and volume

B.1.a) Here is the plot of the function

$$f[x] = 1 + 2\,x + \sin[5\,x]$$

for $a \leq x \leq b$ with $a = 1$ and $b = 3$ shown with some boxes.

```
In[1]:=
  Clear[x,y,f,points,iterations,jump,n,box,boxes,k]
  a = 1; b = 3; f[x_] = 1 + 2 x + Sin[5 x];
  fplot = Plot[f[x],{x,a,b},
  PlotStyle->{{Blue,Thickness[0.01]}},
  AxesLabel->{"x",""},DisplayFunction->Identity];
  iterations[n_] = n; jump[n_] = (b - a)/iterations[n];
  points[n_] := Table[Graphics[{PointSize[0.01],Red,
  Point[{x,0}]}],{x,a,b-jump[n],jump[n]}]
  box[n_,x_] := {Graphics[{GrayLevel[0.7],
  Polygon[{{x,0},{x,f[x]},{x + jump[n],f[x]},
  {x + jump[n],0}}]}],Graphics[Line[{{x,0},{x,f[x]},
  {x + jump[n],f[x]},{x + jump[n],0},{x,0}}]]};
  boxes[n_] := Table[box[n,x],{x,a,b-jump[n],jump[n]}];
  AreaStory[n_] := Show[fplot,boxes[n],points[n],
  PlotLabel->"f[x] and boxes", DisplayFunction->Identity];
  Show[AreaStory[30], DisplayFunction->$DisplayFunction];
```

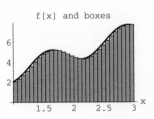

The area measurement $\int_1^3 f[x]\,dx$ is nearly the same as the accumulated measurements of the areas of the boxes. The area of the box whose lower left corner is at the point $\{x, 0\}$ measures out to $f[x] \times$ jump. So the accumulated area of all the boxes measures out to

$$\text{Sum}[f[x]\,\text{jump}, \{x, 1, 3 - \text{jump}, \text{jump}\}].$$

As jump $\rightarrow 0$, these sums close in on

$$\text{Integrate}[f[x], \{x, 1, 3\}] = \int_1^3 f[x]\, dx.$$

Take a look:

In[2]:=
```
Show[AreaStory[100],
DisplayFunction->$DisplayFunction];
```

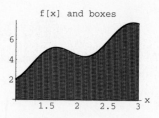

When the jump between consecutive points is so small that it cannot be measured, a lot of folks (especially engineers) say that the jump is dx. In this case,

$$\text{Sum}[f[x]\,\text{jump}, \{x, 1, 3 - \text{jump}, \text{jump}\}] = \text{Sum}[f[x]\,dx, \{x, 1, 3 - dx, dx\}].$$

For all practical purposes, this is the same as Integrate$[f[x], \{x, 1, 3\}] = \int_1^3 f[x]\, dx$.

So what?

Answer: Look at this:

In[3]:=
```
Clear[bars,jump];
bars[jump_] := Graphics[{GrayLevel[0.3],
Table[Line[{{x,0},{x,f[x]}}],{x,a,b,jump}]}];
Show[fplot, bars[(b - a)/60],
DisplayFunction->$DisplayFunction];
```

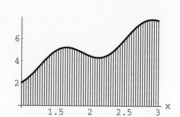

Put in more bars:

In[4]:=
```
Show[fplot, bars[(b - a)/120],
DisplayFunction->$DisplayFunction];
```

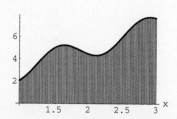

Put in more bars if you like. The upshot is this: If for each x in the plotting interval, you run a bar from $\{x, 0\}$ to $\{x, f[x]\}$ and think of it as a box of height $f[x]$ and of width dx, then the area under the plotted part of the $f[x]$ curve and over the interval from $x = a$ to $x = b$ is

$$\text{Sum}[f[x]dx, \{x, a, b - dx, dx\}]$$

and this is $\int_a^b f[x]\, dx$. Now you have a new vision of integration.

B.1.b.i) Here's a solid.

In[5]:=
```
a = 1; b = 6;
Clear[rad,t,y]
rad[y_] = (y - a) (b - y)/2;
CMView = {2.7, 1.6, 1.2};
solid = ParametricPlot3D[
{rad[y] Cos[t]},y, rad[y] Sin[t]},{y,a,b},{t,0,Pi},
ViewPoint->CMView,Boxed->False,Axes->None];
```

Paint a y-axis through it or next to it and fix measurements a and b as indicated below:

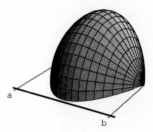

In[6]:=
```
h = 0.2; x = 3.3;
yaxis = {Graphics3D[
{Thickness[0.008],Line[{{x,a-h,0},
{x,b+h,0}}]}],Graphics3D[{GrayLevel[0.5],
Line[{{0,a,0},{x+h,a,0}}]}],
Graphics3D[{GrayLevel[0.5],
Line[{{0,b,0},{x+h,b,0}}]}],
Graphics3D[Text["a",{x + 1.3 h,a,0}]],
Graphics3D[Text["b",{x + 1.3 h,b,0}]]};
solidwithaxis = Show[solid,yaxis,PlotRange->All];
```

The y-axis is the thick line in front.

Now take any y with $a \le y \le b$ and slice the solid with a plane perpendicular to the y-axis that cuts the y-axis at y:

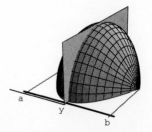

In[7]:=
```
Clear[slicer,y]; xx = 3.5; z = 3.2;
slicer[y_] := {Graphics3D[Polygon[{{xx,y,0},
{-xx,y,0},{-xx,y,z},{xx,y,z}}]],
Graphics3D[Line[{{xx,0,0},{xx,y,0}}]],
Graphics3D[Text["y",{xx + 0.4,y,0}]]};
Show[solidwithaxis,slicer[3.5],
Boxed->False,ViewPoint->CMView];
```

You can slice at any y between a and b: In the plot above, $a = 1.0$ and $b = 6.0$.

In[8]:=
```
Show[solidwithaxis,slicer[2.5],
Boxed->False,ViewPoint->CMView];
```

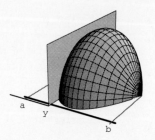

For each y with $a \le y \le b$, agree that Area[y] is the area measurement of the cross section of the solid sliced off by one of these planes.

Explain why the volume of the solid measures out to $\int_a^b$ Area[y] dy.

Answer: Look at a cross section sliced off by one of these planes:

In[9]:=
```
yslice = 3.4; h = 0.1;
slice = ParametricPlot3D[
{rad[y] Cos[t],y, rad[y] Sin[t]},
{y,yslice - h,yslice + h},
{t,0,Pi},DisplayFunction->Identity];
Show[slice,yaxis,slicer[yslice],
ViewPoint->CMView,Boxed->False,
Axes->None,PlotRange->All,
DisplayFunction->$DisplayFunction];
```

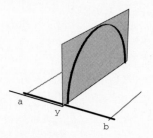

The area of this cross section is Area[y]. When you imagine that this cross section has thickness dy, this tells you that the volume of a slice measures out to Area[y] dy cubic units. As y advances from a to b, the volumes of the slices accumulate to give

$$\text{Sum[Area}[y]\, dy, \{y, a, b - dy, dy\}] = \int_a^b \text{Area}[y]\, dy$$

for the measurement in cubic units of the whole solid. For the solid depicted here, the slices are semicircles of varying radiuses. If you set a y between a and b, the radius is:

In[10]:=
```
rad[y]
```

Out[10]=
$$\frac{(6 - y) \ (-1 + y)}{2}$$

So:

In[11]:=
```
Clear[Area,y]
Area[y_] = Pi rad[y]^2/2
```

Out[11]=

$$\frac{Pi\ (6 - y)^2\ (-1 + y)^2}{8}$$

The volume in cubic units of the solid depicted above is:

In[12]:=
```
NIntegrate[Area[y],{y,a,b}]
```

Out[12]=
```
40.9062
```

Slicing pays off.

B.1.b.ii) The base of a certain solid is a circle of radius 4. All cross sections perpendicular to a fixed diameter are squares.

Measure the volume of the solid.

Answer: Look down from the top:

In[13]:=
```
radius = 4;
base = Graphics[{Blue,Thickness[0.01],
Circle[{0,0},radius]}];
Show[base,PlotLabel->"Top view",
AspectRatio->Automatic];
```

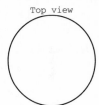

Paint the axes through the center:

In[14]:=
```
Show[base,PlotLabel->"Top view",
Axes->True,AxesLabel->{"x","y"},
AspectRatio->Automatic];
```

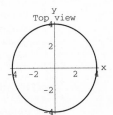

Put in a sample cross section perpendicular to the *x*-axis and to the screen.

In[15]:=
```
Clear[slice,x]
slice[x_] = {Graphics[{Thickness[0.008],Red,
Line[{{x,-Sqrt[radius^2 - x^2]},
{x,Sqrt[radius^2 - x^2]}}]}],
Graphics[Text["x",{x,0.2}]]};
Show[base,slice[0.8],PlotLabel->
"Top view with cross section",
AxesOrigin->{0,0},Axes->Automatic,
AxesLabel->{"x","y"},AspectRatio->Automatic];
```

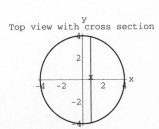

The slice running parallel to the y-axis is the base of the square cross section of the solid. Because the circle's radius is 4, the top curve is $y = \sqrt{16 - x^2}$. So the area measurement Area[x] of the cross section of the solid hitting the x-axis at $\{x, 0\}$ is:

In[16]:=
```
Clear[Area,x]; Area[x_] = (2 Sqrt[radius^2 - x^2])^2
```

Out[16]=

$$4 \ (16 - x^2)$$

The volume in cubic units measures out to $\int_{-4}^{+4}$ Area[x] dx:

In[17]:=
```
NIntegrate[Area[x],{x,-4,4}]
```

Out[17]=
```
341.333
```

Not too hard.

■ B.2) Measurements based on slicing and accumulating: Density and mass

B.2.a.i) Here is the side view of a metal plate with rectangular base. All measurements are in centimeters.

In[18]:=
```
Clear[x,top]
top[x_] = Sin[Pi x/5];
topplot = Plot[top[x],{x,0,5},
DisplayFunction->Identity];
inside = Graphics[{LightSteelBlue,
Polygon[Table[{x,top[x]},{x,0,5,0.2}]]}];
plate = Show[topplot,inside,AspectRatio->Automatic,
DisplayFunction->$DisplayFunction];
```

This plate is 2 centimeters thick.

> Measure the volume of the plate in cubic centimeters.

Answer: The volume is just the measurement of the area (in square centimeters) of the plate as you see it above times the thickness, which is 2 centimeters.

The volume in cubic centimeters is:

In[19]:=
```
volume = 2 NIntegrate[top[x],{x,0,5}]
```

Out[19]=
```
6.3662
```

B.2.a.ii) The metal plate above is made of a varying mixture of copper and aluminum. When you slice the plate along a vertical line x units from the left-hand side, as shown below, you find that the composition is x parts copper and $(5 - x)$ parts aluminum.

In[20]:=
```
x = 2.3; sliceplot = Show[plate,
Graphics[Line[{{x,0},{x,top[x]}}]],
Graphics[Text["x",{x + 0.2,0.2}]]];
```

> Give the density in grams per cubic centimeter of the plate at each point of the vertical line cutting the x-axis at a point x.

Answer: The density of copper is 8.9 grams per cubic centimeter. The density of aluminum is 2.7 grams per cubic centimeter. At all points along this line, the composition is x parts copper and $(5 - x)$ parts aluminum. So in grams per cubic centimeter, the density is:

In[21]:=
```
Clear[density,x]; density[x_] = 8.9 x/5 + 2.7 (5 - x)/5
```

Out[21]=
```
0.54 (5 - x) + 1.78 x
```

B.2.a.iii) The simplest way of measuring the mass of this plate is to put it on a scale and look at the read-out from the scale. If no scale is available, you can measure the mass of the plate by chopping and accumulating.

> Do it.

Answer: Chop the plate into strips:

In[22]:=
```
Clear[iterations,jump,points,strip,strips,x,n]
a = 0; b = 5; iterations[n_] = n;
jump[n_] = (b - a)/iterations[n];
points[n_] := Table[Graphics[{PointSize[0.01],
Red,Point[{x,0}]}],{x,a,b-jump[n],jump[n]}]
strip[n_,x_] := Graphics[Line[{{x,0},{x,top[x]},
{x + jump[n],top[x]},{x + jump[n],0},{x,0}}]];
strips[n_] := Table[strip[n,x],{x,a,b-jump[n],jump[n]}];
StripStory[n_] := Show[plate,strips[n],points[n],
AspectRatio->Automatic,PlotLabel->"plate and strips",
DisplayFunction->Identity];
Show[StripStory[50],DisplayFunction->$DisplayFunction];
```

This plate is well approximated by these narrow strips. Because the density changes very little within each strip, the mass of a strip whose lower left corner is at the point $\{x, 0\}$ is well approximated by

$$\text{density}[x]\, 2\, \text{top}[x]\, \text{jump}.$$

This is density times the volume of the rectangular strip. The accumulated mass of all the strips measures out to

Sum[2 density[x] top[x] jump, $\{x, 0, 5 - \text{jump}, \text{jump}\}$].

When the jump is so small that it cannot be measured, call the jump = dx and realize that for really small jumps like dx the strips approximate the plate excellently:

In[23]:=
```
Show[StripStory[100],
DisplayFunction->$DisplayFunction];
```

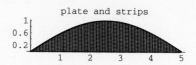
plate and strips

The accumulated mass of the very small strips is:

Sum[2 density[x] top[x] dx, $\{x, 0, 5 - dx, dx\}$].

For all practical purposes, this is the same as:

$$\text{Integrate}[2\,\text{density}[x]\,\text{top}[x], \{x, 0, 5\}] = \int_0^5 2\,\text{density}[x]\,\text{top}[x]\,dx :$$

In[24]:=
```
NIntegrate[2 density[x] top[x],{x,0,5}]
```

Out[24]=
```
36.9239
```

About 37 grams. And you didn't have to dirty your hands or strain your back.

■ B.3) Measurements based on approximating and accumulating: Arc length

Calculus&*Mathematica* thanks Professor Andrew Gleason of Harvard University for some especially pertinent remarks on the topic of arc length.

Here is a curve:

In[25]:=
```
Clear[x,y,t]; a = 0; b = 4;
x[t_] = 2 t Cos[t] + 3; y[t_] =  t Sin[t];
curveplot = ParametricPlot[{x[t],y[t]},
{t,a,b},PlotStyle->Thickness[0.01],
AspectRatio->Automatic,
AxesOrigin->{0,0},AxesLabel->{"x","y"}];
```

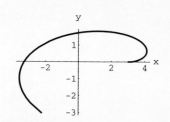

Notice that the plot is in true scale with equal measurements of units along each axis.

B.3.a.i)   Estimate how long this curve is.

Common Sense Answer: The best way to get a fairly accurate estimate is to take a piece of string, lay it along the x-axis, and mark off the units as found on the x-axis directly onto your string. Then take your calibrated string and lay it out on top of the curve and read off how long the string is. Try it. For many practical purposes, the estimate you get will be fine. To make the measurement with great accuracy, go to the next part.

B.3.a.ii) To get a clue about how to make this measurement with great accuracy, look again at the plot of the curve studied in part B.3.a.i) above.

In[26]:=
```
Show[curveplot];
```

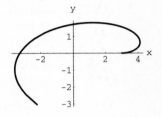

Now look at some Euler fakes of the same curve:

In[27]:=
```
iterations = 15; start = {x[a],y[a]};
jump = (b - a)/iterations;
Clear[t,xx,yy,nextx,nexty,xpoint,ypoint,k]
nextx[{t_,xx_}] := {t,xx} + jump{1, x'[t]}
nexty[{t_,yy_}] := {t,yy} + jump{1, y'[t]}
xpoint[0] = {a,x[a]}; xpoint[k_] :=
N[nextx[xpoint[k-1]]]; ypoint[0] = {a,y[a]};
ypoint[k_] := ypoint[k] = N[nexty[ypoint[k-1]]];
Eulerfake = Graphics[{Red, Thickness[0.01],
Line[Table[{xpoint[k][[2]],ypoint[k][[2]]},
{k,0,iterations}]]}]; Show[curveplot,Eulerfake,
Axes->True,AspectRatio->Automatic,
PlotRange->{{-4,5},{-4,2}},
PlotLabel->iterations"= iterations"];
```

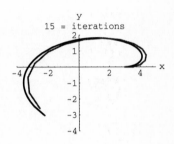

The thinner curve is the plot of Euler's faker.

Step up the iterations:

In[28]:=
```
iterations = 30; start = {x[a],y[a]};
jump = (b - a)/iterations;
Clear[t,xx,yy,nextx,nexty,xpoint,ypoint,k]
nextx[{t_,xx_}] := {t,xx} + jump{1, x'[t]}
nexty[{t_,yy_}] := {t,yy} + jump{1, y'[t]}
xpoint[0] = {a,x[a]}; xpoint[k_] :=
xpoint[k] = N[nextx[xpoint[k-1]]]; ypoint[0] =
{a,y[a]}; ypoint[k_] := ypoint[k] =
N[nexty[ypoint[k-1]]]
Eulerfake = Graphics[{Red, Thickness[0.01],
Line[Table[{xpoint[k][[2]],ypoint[k][[2]]},
{k,0,iterations}]]}]; Show[curveplot,Eulerfake,
Axes->True,PlotRange->{{-4,5},{-4,2}},
PlotLabel->iterations"= iterations"];
```

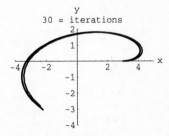

Step up the iterations again:

In[29]:=
```
iterations = 100; start = {x[a],y[a]};
jump = (b - a)/iterations;
Clear[t,xx,yy,nextx,nexty,xpoint,ypoint,k]
nextx[{t_,xx_}] := {t,xx} + jump{1, x'[t]}
nexty[{t_,yy_}] := {t,yy} + jump{1, y'[t]}
xpoint[0] = {a,x[a]}; xpoint[k_] := xpoint[k] =
N[nextx[xpoint[k-1]]]; ypoint[0] = {a,y[a]};
ypoint[k_] := ypoint[k] = N[nexty[ypoint[k-1]]]
Eulerfake = Graphics[{Red,Thickness[0.01],
Line[Table[{xpoint[k][[2]],ypoint[k][[2]]},
{k,0,iterations}]]}]; Show[curveplot,Eulerfake,
Axes->True,AxesLabel->{"x","y"},
PlotRange->{{-4,5},{-4,2}},
PlotLabel->iterations"= iterations"];
```

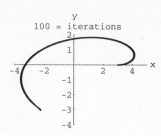

This is just like adjusting your string so that the string is right on top of the curve.

> Explain how these Euler fakers tell you that the exact length of this curve is
> measured by
>
> $$\int_a^b \sqrt{x'[t]^2 + y'[t]^2}\, dt,$$
>
> so that:
>
> *In[30]:=*
> ```
> NIntegrate[Sqrt[x'[t]^2 + y'[t]^2],{t,a,b}]
> ```
>
> *Out[30]=*
>   13.3765
>
> is a high-quality estimate of its length.

Answer: The faker works by generating points:

In[31]:=
```
iterations = 15; start = {x[a],y[a]};
jump = (b - a)/iterations;
Clear[t,xx,yy,nextx,nexty,xpoint,ypoint,k]
nextx[{t_,xx_}] := {t,xx} + jump{1, x'[t]}
nexty[{t_,yy_}] := {t,yy} + jump{1, y'[t]}
xpoint[0] = {a,x[a]}; xpoint[k_] := xpoint[k] =
N[nextx[xpoint[k-1]]]; ypoint[0] = {a,y[a]};
ypoint[k_] := ypoint[k] = N[nexty[ypoint[k-1]]];
Eulerpoints = Table[Graphics[{Red,PointSize[0.02],
Point[{xpoint[k][[2]],ypoint[k][[2]]}]}],{k,0,iterations}];
Show[curveplot,Eulerpoints,Axes->True,
AspectRatio->Automatic,PlotRange->{{-4,5},{-4,2}},
PlotLabel->iterations"= iterations"];
```

and connecting the points with line segments.

In[32]:=
```
Eulerfake = Graphics[{Red,Thickness[0.01],
  Line[Table[{xpoint[k][[2]],
  ypoint[k][[2]]},{k,0,iterations}]]}];
Show[curveplot,Eulerpoints,Eulerfake,
  Axes->True,AspectRatio->Automatic,
  PlotRange->{{-4,5},{-4,2}},
  PlotLabel->iterations"= iterations"];
```

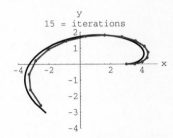

When you go with a big iteration number, the Euler faker gives a great approximation of the curve. So when you go with a big iteration number, the sum of the lengths of the individual line segments gives you a great approximation of the length. If $\{xx, yy\}$ is one of the Euler points corresponding to the parameter t, then $\{xx + \text{jump}\, x'[t], yy + \text{jump}\, y'[t]\}$ is the next Euler point. The length of the Euler line segment connecting these points is

$$\sqrt{(\text{jump}\, x'[t])^2 + (\text{jump}\, y'[t])^2} = \sqrt{\text{jump}^2\,(x'[t]^2 + y'[t]^2)}$$
$$= \text{jump}\,\sqrt{(x'[t]^2 + y'[t]^2)}$$
$$= \text{jump}\,\sqrt{x'[t]^2 + y'[t]^2}.$$

The sum of these lengths is

$$\text{Sum}\left[\sqrt{x'[t]^2 + y'[t]^2}\,\text{jump}, \{t, a, b - \text{jump}, \text{jump}\}\right].$$

As the iteration number gets really big, the jump $\to 0$ and two wonderful things happen:

$\to$ The Euler fakers coalesce with the curve.

$\to$ The lengths of the Euler fakers,

$$\text{Sum}\left[\sqrt{x'[t]^2 + y'[t]^2}\,\text{jump}, \{t, a, b - \text{jump}, \text{jump}\}\right],$$

become

$$\text{Integrate}[\sqrt{x'[t]^2 + y'[t]^2}\,\{t, a, b\}] = \int_a^b \sqrt{x'[t]^2 + y'[t]^2}\,dt.$$

Try it:

In[33]:=
```
iterations = 100;
jump = N[(b - a)/iterations];
EulerEstimate = Sum[Sqrt[x'[t]^2 +  y'[t]^2] jump,{t,a,b - jump,jump}]
```

Out[33]=
```
13.3008
```

In[34]:=
```
iterations = 500; jump = N[(b - a)/iterations];
EulerEstimate = Sum[Sqrt[x'[t]^2 + y'[t]^2] jump,{t,a,b - jump,jump}]
```
Out[34]=
```
13.3612
```

In[35]:=
```
greatEstimate = NIntegrate[Sqrt[x'[t]^2 + y'[t]^2],
{t,a,b}]
```
Out[35]=
```
13.3765
```

This is why the length of the curve is measured by the integral

$$\int_a^b \sqrt{x'[t]^2 + y'[t]^2}\, dt.$$

B.3.b.i) Here is a plot of $f[x] = 1.7 + 0.1\, e^{-x^2/3}$ for $a \le x \le b$ with $a = 2$ and $b = 4$:

In[36]:=
```
Clear[f,x]; a = 2; b = 4;
f[x_] = 1.7 + 0.1 E^((-x^2)/3);
curveplot = Plot[f[x],{x,a,b},
AxesOrigin->{a,f[a]},PlotStyle->
{{Blue,Thickness[0.015]}},
AxesLabel->{"x",""}];
```
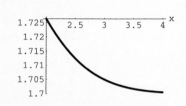

Estimate the length of this curve.

Answer: Take a close look at the numbers on the vertical axis and note that $f[x]$ changes very little as x advances from a to b. For most practical purposes, the plot is that of a horizontal line:

In[37]:=
```
Show[curveplot,PlotRange->{0,3}];
```
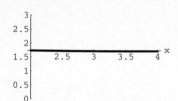

A pretty good estimate is that the curve is about two units long.

B.3.b.ii)   Explain why the integral

$$\int_a^b \sqrt{1 + f'[x]^2}\, dx$$

measures the exact length of the curve in part B.3.b.i).

Answer: Parameterize the curve by $x[t] = t$ and $y[t] = f[t]$ with $a \le t \le b$. The integral

$$\int_a^b \sqrt{x'[t]^2 + y'[t]^2} \, dt$$

measures the length of the curve. But

$$x'[t] = 1$$

and

$$y'[t] = f'[t];$$

so

$$\int_a^b \sqrt{x'[t]^2 + y'[t]^2} \, dt = \int_a^b \sqrt{1^2 + f'[t]^2} \, dt = \int_a^b \sqrt{1 + f'[x]^2} \, dx$$

also measures the length of the curve. Try it:

In[38]:=
```
NIntegrate[Sqrt[1 + f'[x]^2],{x,a,b}]
```

Out[38]=
```
2.00027
```

The eyeball estimate in part B.3.b.i) above was accurate to three decimals. Love those eyes.

B.3.c) Here's a curve in three dimensions:

In[39]:=
```
h = 5; spacer = h/10; threedims = Graphics3D[
{{Blue,Line[{{-h,0,0},{h,0,0}}]},
Text["x",{h + spacer,0,0}],{Blue,Line[
{{0,-h,0},{0,h,0}}]},Text["y",{0, h + spacer,0}],
{Blue,Line[{{0,0,-1},{0,0,h}}]},Text["z",
{0,0,h + spacer}]}]; CMView = {2.7, 1.6, 1.2};
Clear[x,y,z,t]; a = 0; b = 4;
x[t_] = 2 t Sin[t] + 3; y[t_] = t Cos[2 t];
z[t_] = t; curveplot = ParametricPlot3D[
Evaluate[{x[t],y[t],z[t]}],{t,a,b},
DisplayFunction->Identity];
Show[threedims,curveplot,PlotRange->All,
ViewPoint->CMView,Boxed->False,
DisplayFunction->$DisplayFunction];
```

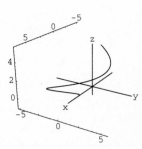

> Measure its length.

Answer: The exact length of this curve is measured by

$$\int_a^b \sqrt{x'[t]^2 + y'[t]^2 + z'[t]^2} \, dt.$$

A very high-quality estimate is:

In[40]:=
```
NIntegrate[Sqrt[x'[t]^2 + y'[t]^2 + z'[t]^2],{t,a,b}]
```

Out[40]=
```
18.7981
```

■ B.4) Measurements based on the fundamental formula: Accumulated growth

The fundamental formula of calculus says

$$\int_a^t f'[x]\,dx = f[t] - f[a].$$

This says that if $f[x]$ is growing at a rate of $f'[x]$ per unit on the x-axis, then its accumulated growth as x advances from a to t is $\int_a^t f'[x]\,dx$. In this interpretation, the integral measures accumulated growth.

B.4.a) You are manager of The C&M Catfish Farm in Tupelo, Mississippi. With time t measured in days from the beginning of the week, you are going to harvest catfish at the rate of

$$r[t] = 1200 + 600\,\sin[2.8\,t] \text{ per day}.$$

> Measure the total catfish harvest for the week.
>
> Measure the total catfish harvest for the first two days.

Answer: As the week progresses, the week's harvest grows at a rate of

$$r[t] = 1200 + 600\,\sin[1.8\,t] \text{ catfish per day}.$$

So the harvest for the week is given by

$$\int_0^7 (1200 + 600\,\sin[1.8\,t])\,dt :$$

In[41]:=
```
Clear[r,t]
r[t_] = 1200 + 600 Sin[1.8 t]; Integrate[r[t],{t,0,7}]
```

Out[41]=
```
8400.19
```

About 8400 catfish for the week. For the first two days:

In[42]:=
```
Clear[t]; Integrate[r[t],{t,0,2}]
```

Out[42]=
```
3032.25
```

The harvest day-by-day:

In[43]:=
```
Clear[day]
Table[{day + 1,Integrate[r[t],{t,day,day + 1}]},{day,0,6}]
```

Out[43]=
```
{{1, 1609.07}, {2, 1423.19}, {3, 689.516}, {4, 1208.78},
 {5, 1706.49}, {6, 961.067}, {7, 802.079}}
```

B.4.b.i) You are moving along the x-axis advancing from $\{1,0\}$ and stopping at $\{10,0\}$. All the while, you are moving in the presence of an electric field emanating from $\{-0.25, 0\}$ that is growing weaker as you advance. In fact, at a point $\{x, 0\}$, the instantaneous rate of voltage drop per unit on the x-axis is

$$\frac{150}{(x + 0.25)^2} \text{ volts per } x\text{-unit.}$$

If at $\{1, 0\}$ the strength of the electricity measures out to 120 volts, then what does the strength of the electricity measure out to at the end of your trip?

Answer: You know:

In[44]:=
```
Clear[volts,voltsprime]; volts[1] = 120
```

Out[44]=
```
120
```

In[45]:=
```
voltsprime[x_] = -150/(x + 0.25)^2
```

Out[45]=
$$\frac{-150}{(0.25 + x)^2}$$

The fundamental formula tells you that

$$\text{volts}[10] = \text{volts}[1] + \int_1^{10} \text{volts}'[x]\, dx :$$

In[46]:=
```
Clear[x]
volts[1] + NIntegrate[voltsprime[x],{x,1,10}]
```

Out[46]=
```
14.6341
```

As you advance from $\{1, 0\}$ to $\{10, 0\}$, the voltage drops from 120 to 14.6.

B.4.b.ii) Continue with the same set-up as in part B.4.b.i). Continue your trip on the x-axis beyond $\{10, 0\}$.

Find a number t^* so that when you go beyond $\{t^*, 0\}$, you will be sure that the voltage is smaller than 5 volts.

Answer:

In[47]:=
```
Clear[volts,voltsprime]
volts[1] = 120
```
Out[47]=
```
120
```

In[48]:=
```
voltsprime[x_] = -150/(x + 0.25)^2
```
Out[48]=
$$\frac{-150}{(0.25 + x)^2}$$

The fundamental formula tells you that the voltage at position $\{t, 0\}$ measures out to:

$$\text{volts}[t] = \text{volts}[1] + \int_1^t \text{volts}'[x]\, dx :$$

In[49]:=
```
Clear[x,t]
volts[t_] = volts[1] + Integrate[voltsprime[x],{x,1,t}]
```
Out[49]=
$$0. + \frac{150}{0.25 + t}$$

Take a look at a plot:

In[50]:=
```
Plot[{5,volts[t]},{t,1,50},
PlotStyle->{{Red},{Thickness[0.01]}},
AxesLabel->{"t","volts"}];
```

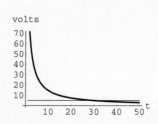

To get below 5 volts, it looks like $t^* = 30$ will do:

In[51]:=
```
Solve[5 == volts[t]]
```
Out[51]=
```
{{t -> 29.75}}
```

Actually, you can get by with $t^* = 29.75$.

Tutorials

■ T.1) Volumes of solids

T.1.a) Here's a surface whose top skin is defined by

$$z = f[x,y] = 4\,\cos[0.5\,x + y]^2$$

as x and y run over the rectangle $-3 \leq x \leq 3$ and $-2 \leq y \leq 2$ in the xy-plane:

In[1]:=
```
Clear[f,x,y]
f[x_,y_] = 4 Cos[0.5 x + y]^2;
a = 3; b = 2; Clear[x,y]
surface = Plot3D[f[x,y],{x,-a,a},{y,-b,b},
DisplayFunction->Identity]; spacer = 0.3;
threedims = Graphics3D[
{{Blue,Line[{{0,0,0},{a + 1,0,0}}]},
Text["x",{a + 1 + spacer,0,0}],
{Blue,Line[{{0,0,0},{0,b + 0.5,0}}]},
Text["y",{0,b + 0.5 + spacer,0}],
{Blue,Line[{{0,0,0},{0,0,5}}]},
Text["z",{0,0,5 + spacer}]}];
CMView = {2.7,1.6,1.2};
plot = Show[surface,threedims,
ViewPoint->CMView,PlotRange->All,
DisplayFunction->$DisplayFunction];
```

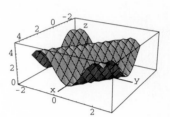

> Measure the volume of the solid whose top skin is this surface and whose bottom skin is everything in the xy-plane that is directly below this surface.

Answer: This is cake once you get the taste of it. Cut the solid with a cutting plane parallel to the xz-plane, hitting the y-axis at $\{0,y,0\}$:

In[2]:=
```
Clear[cutplane,y]; h = 3.5;
cutplane[y_] := {Graphics3D[Polygon[{{-a,y,4.5},
{a + 0.5,y,4.5},{a + 0.5,y,0},{-a,y,0}}]],
Graphics3D[Line[{{a + 0.5,0,0},{a + 0.5,y,0}}]],
Text["y",{a + 0.75,y/2,0}]};
Show[plot,cutplane[1],
Boxed->False,ViewPoint->CMView];
```

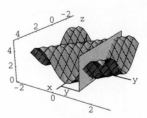

For a given y, the area measurement of this slice is $\int_{-a}^{a} f[x,y]\,dx$ because x runs from -3 to 3:

In[3]:=
```
Clear[Area,y]
Area[y_] = Integrate[f[x,y],{x,-a,a}]
```

Out[3]=

-4. (-1.5 + y) + 4. (1.5 + y) - 2. Sin[2 (-1.5 + y)] + 2. Sin[2 (1.5 + y)]

NIntegrate will not work here because this integrand contains the symbol y.

The slices started with $y = -b$ and stopped with $y = b$ (with $b = 2$); so the volume in cubic units is measured by $\int_{-b}^{b} \text{Area}[y]\,dy$:

In[4]:=
```
volume = NIntegrate[Area[y],{y,-b,b}]
```

Out[4]=
47.5728

Slicing with planes parallel to the yz-plane works, too. If you do this, you get:

In[5]:=
```
Clear[Area,x]
Area[x_] = Integrate[f[x,y],{y,-b,b}];
volume = NIntegrate[Area[x],{x,-a,a}]
```

Out[5]=
47.5728

Either way is calculationally (and politically) correct. But slicing with planes parallel to the xy-plane would result in a calculational nightmare.

T.1.b) Here is the curve $y = 3\,e^{-0.5x}$ plotted for $0 \leq x \leq 5$:

In[6]:=
```
Clear[x,f]
f[x_] = 3 E^(-0.5 x);
Plot[f[x],{x,0,5},
PlotStyle->{{Blue,Thickness[0.015]}},
AxesLabel->{"x","y"}];
```

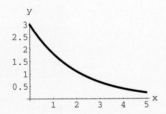

Here is the same curve plotted on the xy-plane in three dimensions:

In[7]:=
```
curve = ParametricPlot3D[{x,f[x],0},{x,0,5},
DisplayFunction->Identity]; spacer = 0.5;
threedims = Graphics3D[
{{Blue,Line[{{0,0,0},{5 + 1,0,0}}]},
Text["x",{5 + 1 + spacer,0,0}],
{Blue,Line[{{0,0,0},{0,3.5 ,0}}]},
Text["y",{0,3.5 + spacer,0}],
{Blue,Line[{{0,0,0},{0,0,3.5 }}]},
Text["z",{0,0,3.5 + spacer}]}];
CMView = {2.7, 1.6, 1.2};
curveplot = Show[curve,threedims,
ViewPoint->CMView,PlotRange->All,Boxed->False,
DisplayFunction->$DisplayFunction];
```

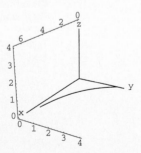

Here's what you get when you rotate this curve about the x-axis:

In[8]:=
```
Clear[t,surfaceplotter];
surfaceplotter[x_,t_] =
{x,f[x] Cos[t],f[x] Sin[t]};
surface = ParametricPlot3D[surfaceplotter[x,t],
{x,0,5},{t,0,2 Pi},DisplayFunction->Identity];
surfaceplot = Show[surface,curveplot,
ViewPoint->CMView,PlotRange->All,Boxed->False,
DisplayFunction->$DisplayFunction];
```

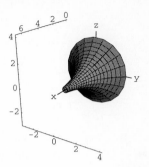

Rad plot.

> Measure the volume of everything inside this surface.

Answer: You can see from the plot that when you slice this solid with a plane parallel to the yz-plane hitting the x-axis at $\{x,0,0\}$, you get a circular disk whose radius is $f[x]$. For a given x, the area measurement of this slice is $\pi\,f[x]^2$.

In[9]:=
```
Clear[x,Area]; Area[x_] = Pi f[x]^2
```

Out[9]=
$$\frac{9\,\text{Pi}}{\text{E}^{1.\,\text{x}}}$$

You started slicing at $x = 0$ and stopped slicing at $x = 5$, so the total volume measurement in cubic units is $\int_0^5 \text{Area}[x]\,dx$:

In[10]:=
```
NIntegrate[Area[x],{x,0,5}]
```

Out[10]=
```
28.0838
```

That was easy.

■ T.2) Linear dimension: Volume and area

The volume measurement, $V[r]$, and the surface area measurement, $S[r]$, of a sphere of radius r are given by

$$V[r] = \frac{4\,\pi\,r^3}{3}$$

and

$$S[r] = 4\,\pi\,r^2.$$

This says that $V[r]$ is proportional to r^3 and $S[r]$ is proportional to r^2. For other three-dimensional objects, the formulas for volume and surface area are not so easy to come by, but the idea of proportionality survives.

Here is the idea: A linear dimension of a given solid or shape is any length between specified locations on the solid. The radius of a sphere or the radius of a circle is a linear dimension. The total length of a solid, the total width, or the total height of a solid are all examples of linear dimensions. The diameter of the finger loop on a coffee cup is a linear dimension of the cup. Next take a given shape for a solid. If the shape stays the same but the linear dimensions change, then it is still true that the volume is proportional to the cube of any linear dimension and it is still true that the surface area is proportional to the square of any linear dimension. Here is a shape in the xy-plane:

In[11]:=
```
Clear[x,y,t]
x[t_] = 4 t (2 - t) E^(t/4);
y[t_] = 2 - Sin[Pi t]; a = 0; b = 2;
plot1 = ParametricPlot[{x[t],y[t]},{t,a,b},
Axes->None,PlotRange->{{0,17},{0,10}},
AspectRatio->Automatic,
PlotStyle->{{Red,Thickness[0.007]}}];
```

Here is the same shape with all its linear dimensions increased by a factor of 3:

In[12]:=
```
plot2 = ParametricPlot[{3 x[t],3 y[t]},{t,a,b},
Axes->None,PlotRange->{{0,17},{0,10}},
AspectRatio->Automatic,
PlotStyle->{{Red,Thickness[0.007]}}];
```

Here they are together:

In[13]:=
```
Show[plot1,plot2,AspectRatio->Automatic];
```

The area measurement of the larger region is $3^2 = 9$ times the area of the smaller region. The idea of linear dimension leads to some intriguing biological implications. A giant mouse with linear dimension 10 times larger than the usual mouse would not be viable because the volume of its body would be larger than the volume of the usual mouse by a factor of 10^3, but the surface area of some of its critical supporting organs, like lungs, intestines, and skin, would be larger only by a factor of 10^2. That big mouse would be hungry and out of breath at all times! Similarly, there will never be a 12-foot-tall basketball player at Indiana or even at Duke. The approximate size of an adult mammal is dictated by its shape! The same common sense applies to buildings and other structures. An architect or engineer does not design a 200-foot-tall building by taking a proven design for a 20-foot-tall building and multiplying all the linear dimensions by 10.

T.2.a) A crystal grows in such a way that all the linear dimensions increase by 25%.

> What are the percentages of the increases in the surface area and the volume?

Answer: The percentage increase in surface area is:

In[14]:=
```
100 (1.25)^2 - 100
```
Out[14]=
```
56.25
```

The surface area increases by about 56%. The percentage increase in volume is:

In[15]:=
```
100 (1.25)^3 - 100
```
Out[15]=
```
95.3125
```

So an increase of the linear dimensions by 1/4 nearly doubles the volume.

T.2.b) A cone is determined by a flat base and a peak:

In[16]:=
```
peak = {0,2,4}; peakplotter = Graphics3D[
{Red,PointSize[0.03],Point[peak]}];
Clear[baseplotter,r,t]
baseplotter[r_,t_] = {r 2 Cos[t],r 3 Sin[t],0};
base = ParametricPlot3D[baseplotter[r,t],
{r,0,1},{t,0, 2 Pi},PlotPoints->{2,Automatic},
DisplayFunction->Identity]; CMView = {2.7, 1.6, 1.2};
Show[base,peakplotter, ViewPoint->CMView,
DisplayFunction->$DisplayFunction,PlotRange->All,
Boxed->False,Axes->None];
```

The cone is made by running straight line segments from each point on the boundary of the base to the peak. Here are some of the line segments:

In[17]:=
```
segments = Table[Graphics3D[Line[
{baseplotter[1,t],peak}]],{t,0, 2 Pi,Pi/4}];
Show[base,peakplotter,segments,
ViewPoint->CMView,DisplayFunction->
$DisplayFunction,PlotRange->All,
Boxed->False,Axes->None];
```

Here are a lot more of them:

In[18]:=
```
segments = Table[Graphics3D[Line[
{baseplotter[1,t],peak}]],{t,0, 2 Pi,Pi/32}];
Show[base,peakplotter,segments,
ViewPoint->CMView,DisplayFunction->$DisplayFunction,
PlotRange->All,Boxed->False,Axes->None];
```

The base of this cone sits on the xy-plane, and the height is 4. Here is what you get when you slice the cone by a plane parallel to the xy-plane but running s units below the top:

In[19]:=
```
h = 4; s = 1.5;
slicer = Graphics3D[
Polygon[{{-2,-3,h - s},{-2,3,h - s},
{2,3,h - s},{2,-3,h - s}}]];
Show[base,peakplotter,segments,slicer,
ViewPoint->CMView,DisplayFunction->
$DisplayFunction,PlotRange->All,
Boxed->False,Axes->None];
```

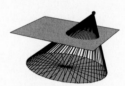

This slicing plane is parallel to the base, and the slicing plane is running s units below the peak. Because the surface of the cone is composed of straight line segments, the linear dimensions of the part of the slicing plane that is inside the cone are all proportional to s.

Use this fact to come up with a formula for the volume inside a cone. Use your formula to measure the volume inside the specific cone plotted above.

Answer: The linear dimensions of this slice are proportional to s. So the area measurement of the cross section sliced off is proportional to s^2. As a result, the area measurement $A[s]$ of a cross section parallel to the base running s units below the peak is given by $A[s] = K s^2$ where K is a constant of proportionality. To find K, notice that when $s = h$, the slicing plane is concurrent with the base:

In[20]:=
```
h = 4; s = h;
slicer = Graphics3D[
Polygon[{{-2,-3,h - s},{-2,3,h - s},
{2,3,h - s},{2,-3,h - s}}]];
Show[base,peakplotter,segments,slicer,
ViewPoint->CMView,DisplayFunction->
$DisplayFunction,PlotRange->All,
Boxed->False,Axes->None];
```

This gives away everything! It tells you that if B stands for the area measurement of the base, then you get

$$B = A[h] = K\,h^2.$$

Divide the equation by h^2 to see that

$$K = \frac{B}{h^2}.$$

Remembering that

$$A[s] = K\,s^2,$$

this gives you:

In[21]:=
```
Clear[A,B,h,s]
A[s_] = (B/h^2) s^2
```
Out[21]=
$$\frac{B\ s^2}{h^2}$$

Because you started slicing with $s = 0$ and stopped with $s = h$, the total volume inside the cone measures out to $\int_0^h A[s]\,ds$:

In[22]:=
```
Integrate[A[s],{s,0,h}]
```
Out[22]=
$$\frac{B\ h}{3}$$

There it is! This formula is beautiful for its pure simplicity. No matter how irregular the planar base, the cone has volume measurement equal to one third times the product of the area measurement of the base and the height measurement. To measure the volume inside the cone plotted above, remember:

In[23]:=
```
h = 4; baseplotter[r,t]
```
Out[23]=
```
{2 r Cos[t], 3 r Sin[t], 0}
```

Here $0 \le r \le 1$ and $0 \le t \le 2\pi$. This plots out as everything inside and on the ellipse

$$\left(\frac{x}{2}\right)^2 + \left(\frac{y}{3}\right)^2 = 1.$$

The area measurement of this is:

In[24]:=
```
B = 2 3 Pi
```
Out[24]=
```
6 Pi
```

So the volume of the cone plotted above measures out to:

In[25]:=
```
B h/3
```
Out[25]=
```
8 Pi
```

in cubic units. And you're out of here.

Give It a Try

Experience with the starred (⋆) problems will be especially beneficial for understanding later lessons.

■ G.1) Using the tools: Measurements of accumulation⋆

G.1.a.i) Calculus&*Mathematica* thanks the farmers who gather for coffee at Gary's Restaurant in Homer, Illinois for some pointers on this problem.

When you are fresh, you find that you can harvest 200 bushels of corn per hour. But as the day wears on, your efficiency is somewhat decreased. In fact, after t hours from the beginning of the day, you find that you are harvesting at a rate of:

In[1]:=
```
Clear[harvestrate,t]
harvestrate[t_] = 200 E^(-0.05 t)
```
Out[1]=
```
  200
 _____
 0.05 t
E
```

bushels per hour. After eight hours in the field, you are so worn out that you are harvesting at a rate of only:

In[2]:=
```
harvestrate[8]
```
Out[2]=
```
134.064
```

bushels per hour.

> Measure your total harvest when you start fresh and work for eight consecutive hours.

G.1.a.ii)
> Estimate the earliest time t that your average harvest rate falls below 180 bushels per hour.

G.1.a.iii) You look at the results from part G.1.a.ii) and decide to work for four consecutive hours, take a big beverage and fried chicken break, and then return to your work for another four hours, starting fresh.

> How many bushels of corn will you harvest this way?
>
> How many more bushels will you harvest this way than you would harvest by working straight through?

G.1.b.i) Water flows into a huge tank at a rate of $g[t]$ gallons/hour. At time $t = 0$, the tank contains G gallons of water.

> Give a formula for the number of gallons of water in the tank at any time $t \geq 0$ before the tank is full.

G.1.b.ii) At time $t = 0$, a 207-gallon tank contains 100 gallons of old water. New water flows in at a rate of

$$g[t] = \frac{160\,t}{(0.5 + t)^2} \text{ gallons/hour.}$$

> Approximately how many hours pass before the tank overflows?

G.1.b.iii) At time $t = 0$, a 207-gallon tank contains 100 gallons of old water. New water flows in at a rate of

$$g[t] = \frac{160\,e^t}{(0.5 + e^t)^2} \text{ gallons/hour.}$$

> Does the tank ever overflow? How do you know?

G.1.b.iv) At time $t = 0$, a 1500-gallon tank contains 1000 gallons of old water. New water flows in at a rate of

$$g[t] = \frac{K\,e^t}{0.5 + e^t} \text{ gallons/hour.}$$

But water is pumped out at a rate of 80 gallons per hour.

> Your job is to set the constant K so that the tank neither overflows nor runs dry for the next 100 days ($= 2400$ hours).

G.1.c.i) You are moving along the x-axis advancing from $\{2,0\}$ and stopping at $\{12,0\}$. All the while, you are moving in the presence of an electric field emanating from $\{-0.5,0\}$ that is growing weaker as you advance. In fact, at a point $\{x,0\}$, the instantaneous rate of voltage drop per unit on the x-axis is $90/\left(x+0.5\right)^2$ volts per x-unit.

> If at $\{2,0\}$, the strength of the electricity measures out to 30 volts, then what does the strength of the electricity measure out to at the end of your trip?

G.1.c.ii) Continue with the same set-up as in part G.1.c.i). Continue your trip on the x-axis beyond $\{12,0\}$.

> Find a number t^* so that when you go beyond $\{t^*,0\}$, you will be sure that the voltage is smaller than 2.5 volts.

■ G.2) Measurements of length, volume, and mass⋆

G.2.a.i) Every school kid has been told that a circle of radius r has circumference $2\pi r$. Here is a parameterization of a circle of radius r:

In[3]:=
```
Clear[x,y,t,r]
{x[t_],y[t_]} = {r Cos[t],r Sin[t]}
```

Out[3]=
```
{r Cos[t], r Sin[t]}
```

To get the whole circle, you run t from 0 to 2π.

> Use this information to confirm that the circumference of a circle of radius r measures out to $2\pi r$ units long.

G.2.a.ii) Every school kid has been told that a circle of radius r encloses an area of πr^2 square units.

> Explain why the following calculation confirms this formula:
>
> *In[4]:=*
> ```
> Clear[x,y,r]
> Solve[x^2 + y^2 == r^2,y]
> ```
> *Out[4]=*
> ```
> {{y -> Sqrt[r  - x ]}, {y -> -Sqrt[r  - x ]}}
> ```
> $$\{\{y \to \text{Sqrt}[r^2 - x^2]\}, \{y \to -\text{Sqrt}[r^2 - x^2]\}\}$$

```
In[5]:=
   areameasurement = 2 Integrate[Sqrt[r^2 - x^2],{x,-r,r}]
Out[5]=
        2        r
   2 r  ArcSin[--------]
                     2
              Sqrt[r ]
```

Mathematica does not know whether r is positive or negative, but you know that r is positive. Clean up the calculation with:

```
In[6]:=
   betterareameasurement = areameasurement/.ArcSin[r/Sqrt[r^2]]->ArcSin[1]
Out[6]=
        2
   Pi r
```

G.2.b.i) All reference books agree that the volume enclosed by a sphere of radius r measures out to $4/3\,\pi\,r^3$ cubic units.

> Where does this formula come from?

G.2.b.ii) A hemispherical bowl of radius r contains water to a depth h.

> Give a formula that can be used to measure the volume of the water in the bowl.

G.2.c) Here is a surface:

```
In[7]:=
   Clear[x,y]
   surface = Plot3D[1 + Sin[x + y],
   {x,0,2 Pi},{y,0,2 Pi},
   DisplayFunction->Identity]; h = 7;
   spacer = h/10; threedims = Graphics3D[
   {{Blue,Line[{{-1,0,0},{h,0,0}}]},
   Text["x",{h + spacer,0,0}],
   {Blue,Line[{{0,-1,0},{0,h ,0}}]},
   Text["y",{0, h  + spacer,0}],
   {Blue,Line[{{0,0,-1},{0,0,h/3 }}]},
   Text["z",{0,0,h/3 + spacer}]}];
   CMView = {2.7, 1.6, 1.2};
   Show[surface,threedims,ViewPoint->CMView,
   PlotRange->All,DisplayFunction->$DisplayFunction];
```

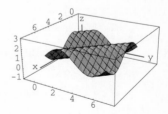

> Measure the volume of the solid whose top skin is this surface and whose bottom skin is everything in the xy-plane that is directly below this surface.

G.2.d.i) Here is the curve

$$y = \frac{8\,x}{1 + x^2} \qquad \text{for } 0 \le x \le 10 :$$

In[8]:=
```
Clear[f,x]
f[x_] = 8 x/(1 + x^2);
ParametricPlot[{x,f[x]},
{x,0,10},AspectRatio->Automatic,
PlotStyle->{{Blue,Thickness[0.015]}},
AxesLabel->{"x","y"}];
```

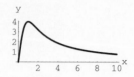

Rotate this curve about the x-axis to form a giant vase:

In[9]:=
```
Clear[t]; vase = ParametricPlot3D[
{x,f[x] Cos[t],f[x] Sin[t]},{x,0,10},{t,0,2 Pi},
DisplayFunction->Identity];
spacer = 0.5; threedims = Graphics3D[
{{Blue,Line[{{0,0,0},{11,0,0}}]},
Text["x",{11 + spacer,0,0}],
{Blue,Line[{{0,0,0},{0,5,0}}]},
Text["y",{0, 5 + spacer,0}],
{Blue,Line[{{0,0,0},{0,0,5 }}]},
Text["z",{0,0,5 + spacer}]}];
CMView = {2.7, 1.6, 1.2};
Show[vase,threedims,ViewPoint->CMView,
PlotRange->All,DisplayFunction->$DisplayFunction];
```

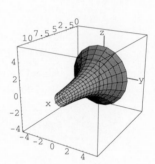

Measure the volume inside this vase.

G.2.d.ii) Here is the curve $x = 2\,y \sin[y]^2$ for $0 \le y \le \pi$:

In[10]:=
```
Clear[f,y]
f[y_] = 2 y Sin[2 y]^2;
ParametricPlot[{f[y],y},{y,0,Pi},
AspectRatio->Automatic,
PlotStyle->{{Blue,Thickness[0.01]}},
AxesLabel->{"x","y"}];
```

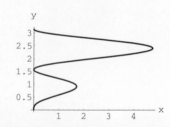

Plot what you get when you rotate this curve about the y-axis and measure the volume enclosed.

G.2.e) A certain cone has a planar base whose area measures out to 50 square units, and the height of this cone is 12 units.

Measure the volume inside the cone.

G.2.f.i) This problem appears only in the electronic version.

G.2.f.ii) Here is the side view of a ceramic plate. All measurements are in centimeters.

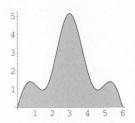

```
In[11]:=
  Clear[x,top]
  top[x_] = x ( 6 - x) (1 + 3 Cos[Pi x/3]^2)/7;
  topplot = Plot[top[x],{x,0,6},
  DisplayFunction->Identity];
  inside = Graphics[{Firebrick,
  Polygon[Table[{x,top[x]},{x,0,6,0.1}]]}];
  plate = Show[inside,topplot,AspectRatio->Automatic,
  DisplayFunction->$DisplayFunction];
```

This plate is 7.3 centimeters thick.

> Measure the volume of the plate in cubic centimeters.

G.2.f.iii) The plate above is made of a varying mixture of ceramics. When you slice the plate along a vertical line x units from the left-hand side, you find that the density of the plate is given by

$$\text{density}[x] = 0.7 + x\ (6 - x)$$

grams per cubic centimeter. The simplest way of measuring the mass of this plate is to put it on a scale and look at the read-out from the scale. If no scale is available, you can measure the mass of the plate with an integral.

> Do it.

■ G.3) Slicing for area measurements

G.3.a.i) Here is a plot of

$$f[x] = \sin[x^2] + 5$$

together with a plot of

$$g[x] = \sin[x^2] + 1$$

for $-3 \le x \le 3$.

```
In[12]:=
  a = -3; b = 3; Clear[f,g,x]
  f[x_] = Sin[x^2] + 5;
  g[x_] = Sin[x^2] + 1;
  plot = Plot[{f[x],g[x]},{x,-3,3},
  PlotStyle->{{Thickness[0.015],Blue},
  {Thickness[0.015],Blue}},
  AspectRatio->Automatic,AxesLabel->{"x",""}];
```

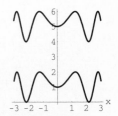

Measuring the area between these two plots is a snap. You just remember that
$f[x] = \sin[x^2] + 5$, $g[x] = \sin[x^2] + 1$, and then you calculate

$$\int_{-3}^{3} (f[x] - g[x])\, dx = \int_{-3}^{3} (5 - 1)\, dx = \int_{-3}^{3} 4\, dx = 24 \text{ square units.}$$

To deepen your understanding of why this integral makes the measurement, look
at this:

In[13]:=
```
Clear[jump,box,boxes,BoxStory,x,n]
jump[n_] = (b - a)/n;
box[n_,x_] := Graphics[{Thickness[0.01],
Line[{{x,g[x]},{x + jump[n],g[x]},
{x + jump[n],f[x]},{x , f[x]},{x,g[x]}}]}];
boxes[n_] :=  Table[box[n,x],{x,a,b-jump[n],jump[n]}];
BoxStory[n_] := Show[plot,boxes[n],AspectRatio->Automatic,
DisplayFunction->Identity]; Show[BoxStory[18],
DisplayFunction->$DisplayFunction];
```

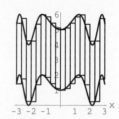

Does each box measure out to the same area as the others?

What happens when you step up the number of boxes?

How do the boxes reveal that the area between the two plotted curves measures
out to 6 times $4 = 24$ square units?

G.3.a.ii) Continuing with the same functions and same endpoints, look at this:

In[14]:=
```
Clear[bars,t,x]; bars[t_] := Graphics[{RGBColor[t,0,1 - t],
Table[Line[{{x,f[x] - t g[x]},{x,(1 - t) g[x]}}],{x,a,b,(b - a)/60}]}];
plots = Plot[{f[x],g[x],f[x] - g[x]},{x,a,b},PlotStyle->
{{Thickness[0.01],Blue},{Thickness[0.01],Blue},{Thickness[0.01],Red}},
DisplayFunction->Identity]; Table[Show[plots,bars[t],PlotRange->All,
DisplayFunction->$DisplayFunction],{t,0,1,0.2}];
```

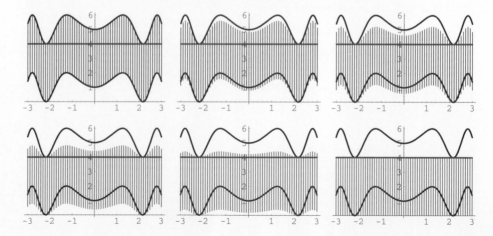

The bars move vertically but not horizontally.

How do the bars confirm to you that the area between the two curves and the area in the rectangle measure out to the same number?

G.3.b.i) Here is a plot of
$$f[x] = \frac{x}{2} + \cos[2\,x]$$
with a plot of
$$g[x] = 1 - 2\,\sin[x]^2$$
for $0 \le x \le 6$.

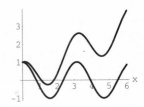

In[15]:=
```
a = 0; b = 6; Clear[f,g,x]
f[x_] = x/2 + Cos[2 x];
g[x_] = 1 - 2 Sin[x]^2;
plot = Plot[{f[x],g[x]},{x,a,b},
PlotStyle->{{Thickness[0.015],Blue},
{Thickness[0.015],Blue}},
AspectRatio->Automatic,AxesLabel->{"x",""}];
```

Look at this:

In[16]:=
```
Clear[bars,t,x];
bars[t_] := Graphics[{RGBColor[t,0,1 - t],
Table[Line[{{x,f[x] - t g[x]},{x,(1 - t) g[x]}}],{x,a,b,(b - a)/60}]}];
plots = Plot[{f[x],g[x],f[x] - g[x]},{x,a,b},PlotStyle->
{{Thickness[0.01],Blue},{Thickness[0.01],Blue},{Thickness[0.01],Red}},
DisplayFunction->Identity]; Table[Show[plots,bars[t],PlotRange->All,
DisplayFunction->$DisplayFunction],{t,0,1,0.2}];
```

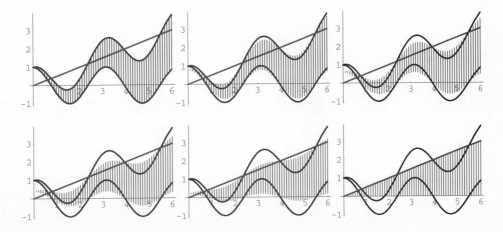

The bars move vertically but not horizontally.

> How do these frames reveal the measurement of the area between the plotted part of the two curves?

G.3.b.ii) Continuing with the same functions and the same endpoints, look at:

In[17]:=
```
Expand[f[x] - g[x],Trig->True]
```
Out[17]=

$$\frac{x}{2}$$

> How does the result allow you to measure easily by hand the area between the two curves without running the graphs?

G.3.c) Make up your own movie that illustrates the same idea that drives parts G.3.a) and G.3.b) above.

G.3.d.i) Here's the circle $x^2 + y^2 = 1$ shown with the ellipse $(x/2)^2 + y^2 = 1$:

In[18]:=
```
circle = Graphics[{RGBColor[0,0,1],
Thickness[0.005],Circle[{0,0},1]}];
ellipse = Graphics[{RGBColor[1,0,0],
Thickness[0.005],Circle[{0,0},{2,1}]}];
Show[circle,ellipse,Axes->True,
AxesLabel->{"x","y"},AspectRatio->Automatic];
```

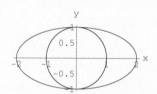

Now look at this:

In[19]:=
```
Clear[bars,t,x]; bars[t_] := Graphics[{RGBColor[t - 1,0,2 - t],
Table[Line[{{-t Cos[x],Sin[x]},{t Cos[x],Sin[x]}}],{x,-Pi/2,Pi/2,Pi/40}]}];
Table[Show[circle,ellipse,bars[t],Axes->True,AxesLabel->{"x","y"},
AspectRatio->Automatic],{t,1,2,0.2}];
```

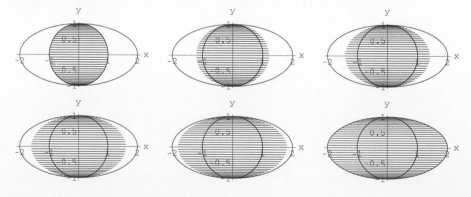

The bars double their lengths from the beginning to the end.

> Given that the area inside the circle $x^2 + y^2 = 1$ measures out to π square units, how do the graphs reveal the measurement of the area inside the ellipse $(x/2)^2 + y^2 = 1$?

G.3.d.ii)
> Given a positive constant a, what does the area inside the ellipse
>
> $$\left(\frac{x}{a}\right)^2 + y^2 = 1$$
>
> measure out to?

G.3.d.iii)
> Given positive constants a and b, what does the area inside the ellipse
>
> $$\left(\frac{x}{a}\right)^2 + \left(\frac{y}{b}\right)^2 = 1$$
>
> measure out to?

■ G.4) Volume measurements for some tubes and horns*

Here is a modest curve shown with the three coordinate axes in three dimensions:

```
In[20]:=
  h = 5; spacer = h/10;
  threedims = Graphics3D[{
  {Blue,Line[{{-h,0,0},{h,0,0}}]},
  Text["x",{h + spacer,0,0}],
  {Blue,Line[{{0,-h,0},{0,h + 1,0}}]},
  Text["y",{0, h + 1 + spacer,0}],
  {Blue,Line[{{0,0,-h/2},{0,0,h + 2}}]},
  Text["z",{0,0,h +2+ spacer}]}];
  CMView = {2.7, 1.6, 1.2};
  Clear[t,curveplotter]
  curveplotter[t_] = {Cos[t],3 t Sin[t],3 t};
  curve = ParametricPlot3D[Evaluate[curveplotter[t]],{t,0,2},
  DisplayFunction->Identity];
  Show[threedims,curve,ViewPoint->CMView,Boxed->False,
  PlotRange->All,DisplayFunction->$DisplayFunction];
```

Here is the same curve shown with a few sample circles parallel to the xy-plane and centered on the curve:

In[21]:=

```
Clear[t]
circle1 = ParametricPlot3D[
Evaluate[curveplotter[0.5] +
2 {Cos[t],Sin[t],0}],
{t,0,2 Pi},DisplayFunction->Identity];
circle2 = ParametricPlot3D[
Evaluate[curveplotter[1] +
2 {Cos[t],Sin[t],0}],
{t,0,2 Pi},DisplayFunction->Identity];
circle3 = ParametricPlot3D[
Evaluate[curveplotter[1.5] +
2 {Cos[t],Sin[t],0}],
{t,0,2 Pi},DisplayFunction->Identity];
Show[threedims,curve,circle1,
circle2,circle3,ViewPoint->CMView,
Boxed->False,PlotRange->All,
DisplayFunction->$DisplayFunction];
```

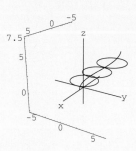

Here's what you get when you make a tube consisting of all such circles:

In[22]:=

```
Clear[tubeplotter,s,t]
tubeplotter[s_,t_] = curveplotter[t] +
2 {Cos[s],Sin[s],0};
tube = ParametricPlot3D[
Evaluate[tubeplotter[s,t]],
{t,0,2},{s,0,2 Pi},
DisplayFunction->Identity];
tubeplot = Show[threedims,tube,
curve,ViewPoint->CMView,
Boxed->False,PlotRange->All,
DisplayFunction->$DisplayFunction];
```

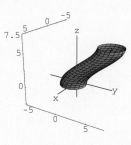

Move each circle in its own plane but so that it is centered on the *z*-axis:

In[23]:=

```
Clear[cylinderplotter,s,t]
cylinderplotter[s_,t_] = {0,0,curveplotter[t][[3]]} + 2 {Cos[s],Sin[s],0}
```

Out[23]=

```
{2 Cos[s], 2 Sin[s], 3 t}
```

Compare:

In[24]:=

```
tubeplotter[s,t]
```

Out[24]=

```
{2 Cos[s] + Cos[t], 2 Sin[s] + 3 t Sin[t], 3 t}
```

Plot the tube:

In[25]:=
```
Clear[s,t]
cylinder = ParametricPlot3D[
Evaluate[cylinderplotter[s,t]],
{t,0,2},{s,0,2 Pi},
PlotPoints->{20,Automatic},
DisplayFunction->Identity];
cylinderplot = Show[threedims,
cylinder,ViewPoint->CMView,
Boxed->False,PlotRange->All,
DisplayFunction->$DisplayFunction];
```

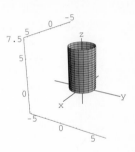

G.4.a.i) The cylinder has a base with radius 2, and the height of the cylinder is 3, so the volume of the cylinder measures out to $2^2\,\pi\,3 = 12\,\pi$ cubic units.

> Explain why the volume of the tube also measures out to $12\,\pi$ cubic units.

G.4.a.ii) Calculus&*Mathematica* thanks former C&M student Jonathan Paetsch for suggesting this part.

> If you go back to the beginning, keeping everything the same but changing curveplotter$[t] = \{\cos[t], 3\,t\,\sin[t], 3\,t\}$ to
>
> $$\text{curveplotter}[t] = \{f[t], g[t], 3\,t\}$$
>
> where $f[t]$ and $g[t]$ are any functions you care to use, then how do you know that the volume of the resulting tube measures out to $12\,\pi$ cubic units?

G.4.b) Look at this plot:

In[26]:=
```
h = 5; spacer = h/10;
threedims = Graphics3D[{{Blue,Line[{{-h,0,0},{h,0,0}}]},
Text["x",{h + spacer,0,0}],{Blue,Line[{{0,-h,0},{0,h + 1,0}}]},
Text["y",{0, h + 1 + spacer,0}],{Blue,Line[{{0,0,-h/2},{0,0,h + 2}}]},
Text["z",{0,0,h + 2 + spacer}]}]; CMView = {2.7, 1.6, 1.2};
```

In[27]:=
```
Clear[s,t,curveplotter,hornplotter,radius]
curveplotter[t_] = {t(4 - t),t,(t^2)/4};
radius[t_] = 0.75 t;
hornplotter[s_,t_] = curveplotter[t] +
radius[t] {Cos[s],0,Sin[s]};
horn = ParametricPlot3D[
Evaluate[hornplotter[s,t]],{t,0,4},
{s,0,2 Pi}, DisplayFunction->Identity];
hornplot = Show[threedims,horn,
ViewPoint->CMView,Boxed->False,PlotRange->All,
DisplayFunction->$DisplayFunction];
```

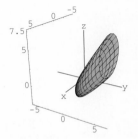

Here's a look from a vantage point on the positive z-axis:

In[28]:=
```
Show[threedims,horn,
ViewPoint->{0,0, 8.000},
Boxed->False,PlotRange->All,
DisplayFunction->$DisplayFunction];
```

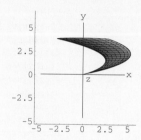

The cross sections of this horn sliced by planes parallel to the xz-plane are circles of varying radiuses.

> Move each circle in its own plane but so that it is centered on the y-axis; and use what you see to measure the volume of this tube.

G.4.c) Start with a base curve specified by

$$\text{curveplotter}[t] = \{t, f[t], g[t]\} \qquad \text{with } a \le t \le b.$$

Build a horn by centering at each point $\{t, f[t], g[t]\}$ a circle of radius $r[t]$ parallel to the yz-plane.

> Write down a formula that everyone can use to measure the volume of this horn.

■ G.5) Work

Factoring polynomials is fun for most math teachers; but factoring polynomials is work for most math students.

We all have our own idea of what work is. What is work for one person often is recreation for another. Physicists have their own technical definition of work: If an object moves along the x-axis from $x = a$ to $x = b$ in the presence of a constant force, then physicists say that force does work in the amount

$$\text{work} = \text{force } (b - a) \ (= \text{force times distance}).$$

A positive force indicates that the push is directed to the right, and a negative force indicates that the push is directed to the left. The question here is how to measure work if the force varies from position to position.

Suppose you have an object moving from $x = 0$ to $x = 6$ on the x-axis under the influence of a force field given by

$$\text{force}[x] = 0.5 + \sin[2\,x]$$

at each point x between $x = 0$ and $x = 6$.

In[29]:=

```
Clear[x,force]
force[x_] = 0.5 + Sin[2 x];
forceplot = Plot[force[x],{x,0,6},
PlotStyle->{Red,Thickness[0.015]}];
```

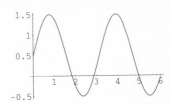

Most of the way, force[x] pushes to the right, but some of the way, force[x] pushes to the left.

G.5.a) Sticking with the same force[x] as above, look at some boxes:

In[30]:=

```
a = 0; b = 6; Clear[jump,points,box,boxes,x,n]
jump[n_] = (b - a)/n; points[n_] :=
Table[Graphics[{PointSize[0.01],Red,
Point[{x,0}]}],{x,a,b-jump[n],jump[n]}]
box[n_,x_] := Graphics[Line[{{x,0},{x,force[x]},
{x + jump[n],force[x]},{x + jump[n],0},{x,0}}]];
boxes[n_] := Table[box[n,x],{x,a,b-jump[n],jump[n]}];
BoxStory[n_] := Show[forceplot,boxes[n],points[n],
AspectRatio->Automatic,DisplayFunction->Identity];
Show[BoxStory[50],DisplayFunction->$DisplayFunction];
```

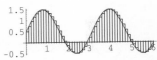

Because force[x] changes very little between each plotted point on the x-axis, the work done by the force in moving from a plotted point $\{x, 0\}$ to the next plotted point $\{x + \text{jump}, 0\}$ is well approximated by force[x] times jump. This is the same as the area of a box whose left side touches the x-axis at $\{x, 0\}$. As a result, the accumulated work in moving all the way from $x = 0$ to $x = 6$ is well approximated by

Sum[force[x] jump, $\{x, 0, 6 - \text{jump}, \text{jump}\}$].

> Use this as a start to try to explain why physicists say that the work done by the force field as the object moves from $x = 0$ to $x = 6$ is
>
> $$\int_0^6 \text{force}[x]\, dx.$$

G.5.b) If you move an object along the x-axis from $x = a$ to $x = b$ in the presence of a variable force given by force[x] at any point between $x = a$ and $x = b$, then $\int_a^b \text{force}[x]\, dx$ measures the amount of work done by the force field. The bigger this measurement is, the more the force field helped you.

> If you are moving an object from $x = 1$ to $x = 7$, which of the following force fields will give you the most help:

```
force1[x_] = (x - 1) ( x - 7)
force2[x_] = (x - 1) ( 7 - x)
force3[x_] = 3 + Sin[2 x]?
```

G.5.c) A cable of length 50 meters and density 4 kilograms per meter hangs over the edge of a skyscraper.

> What work is needed to pull the whole cable to the top of the building?

G.5.d) Physicists measure force in newtons and distance in meters.

> Say why you think physicists measure work in newton-meters (which physicists call a joule).

For your information and enjoyment: power = joules/second. One unit of power is also called a watt. This is where the electric meter on the side of your house gets the measurement of kilowatt hours. A kilowatt hour is

1000 watts times 3600 seconds $= 3.610^6$ (joules/second) seconds $= 3.610^6$ joules.

Every time you pay your power bill, you are paying for units of work done by electricity in keeping your computer running.

G.5.e)
> Write a few words on what you think is the difference between the everyday English language definition of work and the technical definition of work as used by physicists.
>
> To get started, think about this: According to physicists, if there is no change of position, then there is no work. If you must hold a heavy, old 1989 version of the Macintosh II in your arms while you stand in place for one hour, the physicist would say that you did no work.
>
> Do you agree?

■ G.6) Champagne glasses with a logarithmic flare

Here's a function with a logarithmic flare:

In[31]:=
```
Clear[y,a,b,xprofile]
xprofile[y_,a_,b_] = a (Log[y] - Log[(b - y)])
```
Out[31]=
```
a (-Log[b - y] + Log[y])
```

G.6.a.i) How do you know in advance that this function plots out only for $0 < y < b$?

G.6.a.ii) Put

$$\text{xprofile}[y, a, b] = a \, \log \left[\frac{y}{b - y} \right]$$

and explain how you know in advance that the following instruction exhibits the y that makes $\text{xprofile}[y, a, b] = 0$:

In[32]:=
```
Solve[y/(b - y) == 1,y]
```

Out[32]=
```
       b
{{y -> -}}
       2
```

G.6.b) Look at:

In[33]:=
```
a = 0.5; b = 3; c = 0.9;
ParametricPlot[
{{xprofile[y,a,b],y},{-xprofile[y,a,b],y}},
{y,b/2,c b},PlotStyle->{{MidnightBlue,
Thickness[0.015]}},AxesOrigin->{0,b/2},
AxesLabel->{"x","y"}];
```

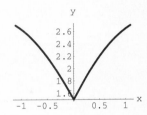

And look at the corresponding measurement

$$\int_{b/2}^{cb} \pi \, \text{xprofile}[y, a, b]^2 \, dy :$$

In[34]:=
```
NIntegrate[Pi xprofile[y,a,b]^2,{y,b/2,c b}]
```

Out[34]=
```
1.13761
```

Next look at:

In[35]:=
```
a = 0.3; b = 8; c = 0.99;
ParametricPlot[
{{xprofile[y,a,b],y},{-xprofile[y,a,b],y}},
{y,b/2,c b},PlotStyle->{{MidnightBlue,
Thickness[0.015]}},AxesOrigin->{0,b/2},
AxesLabel->{"x","y"}];
```

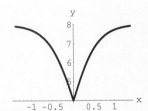

And look at the corresponding measurement

$$\int_{b/2}^{cb} \pi \, \text{xprofile}[y, a, b]^2 \, dy :$$

In[36]:=
```
NIntegrate[Pi xprofile[y,a,b]^2,{y,b/2,c b}]
```

Out[36]=
```
2.98863
```

Again:

In[37]:=
```
a = 2; b = 30; c = 0.999;
ParametricPlot[
{{xprofile[y,a,b],y},{-xprofile[y,a,b],y}},
{y,b/2,c b},PlotStyle->{{MidnightBlue,
Thickness[0.015]}},AxesOrigin->{0,b/2},
AxesLabel->{"x","y"}];
```

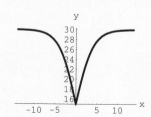

And the corresponding measurement

$$\int_{b/2}^{cb} \pi \, \text{xprofile}[y, a, b]^2 \, dy :$$

In[38]:=
```
NIntegrate[Pi xprofile[y,a,b]^2,{y,b/2,c b}]
```

Out[38]=
```
596.177
```

> What do these measurements reflect?

G.6.c) For each of these choices of $a, b,$ and c, you can think of these curves as the outline of the top part of a champagne glass made by rotating the plotted curve for

$$\frac{b}{2} \leq y \leq cb$$

around the y-axis. A stem and a base will be added later. Here's a sample profile:

In[39]:=
```
a = 1.2; b = 27; c = 0.999;
ParametricPlot[
{{xprofile[y,a,b],y},{-xprofile[y,a,b],y}},
{y,b/2,c b},PlotStyle->{{MidnightBlue,
Thickness[0.015]}},AxesOrigin->{0,b/2},
AxesLabel->{"x","y"}];
```

And the resulting glass without stem or base:

In[40]:=
```
Clear[t]; CMView = {2.7, 1.6, 1.2};
ParametricPlot3D[{Cos[t] xprofile[y,a,b],
Sin[t] xprofile[y,a,b],y},{y,b/2,c b},{t,0,2 Pi},
PlotRange->All,ViewPoint->CMView,
PlotLabel->"champagne glass",
Boxed->False,Axes->None];
```

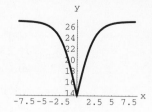

champagne glass

This is a silly champagne glass because the brim of this glass is so wide that the bubbly will spill out from both ends of your lips as you quaff from it.

Now it's time to get creative. Using centimeters as the unit of measurement, adjust the parameters a, b, and c and the curve xprofile$[y, a, b]$ to design the most graceful champagne glass with a capacity of approximately 8-fluid-ounces that you can come up with.

> Plot the profile and the top of your design as above. Throw in a stem and base if you like, but this is not required.

There are 1000 cubic centimeters in a liter, and there are 33.814 fluid ounces in a liter.

■ G.7) The derivative of arc length*

This problem appears only in the electronic version.

■ G.8) Present value of a profit-making scheme

G.8.a.i) You invest $A[0]$ dollars with an investment outfit with the stipulations that:

→ Payments will be made to you or your heirs at a rate of $p[t]$ dollars per year t years from now.

→ If $A[t]$ dollars is the amount in the fund t years from today, then these $A[t]$ dollars are to accrue interest compounded continuously at a rate of $100\,r$ percent where the rate r is held constant.

→ $\lim_{t \to \infty} A[t] = 0$.

(This arrangement is sometimes called a perpetual annuity or perpetuity.)

> Explain why
> $$A'[t] = -p[t] + r\,A[t].$$
> Then explain why
> $$e^{-rt}\,A'[t] - r\,e^{-rt}\,A[t] = -p[t]\,e^{-rt}.$$

G.8.a.ii) Look at:

```
In[41]:=
    Clear[A,t,r,p]; D[E^(- r t) A[t],t]

Out[41]=
      r A[t]      A'[t]
   -(--------) + --------
       r t          r t
      E            E
```

Then integrate both sides of

$$e^{-rt} A'[t] - r\, e^{-rt}\, A[t] = -p[t]\, e^{-rt}$$

from 0 to ∞ to explain why the original investment $A[0]$ is given by

$$A[0] = \int_0^\infty e^{-rt}\, p[t]\, dt.$$

G.8.b) The formula above explains what the financial people call present value. They say that a profit-making scheme that will pay profits at a rate of $p[t]$ dollars per year t years from now has a present value of

$$\int_0^\infty e^{-rt}\, p[t]\, dt$$

where $100\, r$ percent is the projected interest rate (compounded every instant) over the future. So the present value is nothing but the amount of initial investment $A[0]$ that it would take to get payments at the rate of $p[t]$ dollars per year t years from now in the perpetual annuity set-up studied in part G.8.a) above.

You have a profit-making scheme that is projected to pay profits at a rate of

$$p[t] = (100000 + t)\, e^{-t/5}$$

dollars per year t years from now.

Assuming a projected interest rate of 6% compounded every instant, what is the present value of your scheme?

Still assuming a projected interest rate of 6% compounded every instant, how much would you have to plunk down for a perpetual annuity that would pay you at the same rate?

What is the projected total take on this scheme?

How many years would it take for this scheme to play out in the sense that the future take will be next-to-nothing?

■ G.9) Linear dimension*

G.9.a) Here's a shape in the xy-plane:

```
In[42]:=
  Clear[x,y,t]; a = 0; b = 2 Pi;
  x[t_] = Sin[4 t] Cos[t];
  y[t_] = Sin[4 t] Sin[t];
  plot1 = ParametricPlot[{x[t],y[t]},{t,a,b},
  PlotRange->{{-7,7},{-7,7}},
  AspectRatio->Automatic,PlotStyle->
  {{Red,Thickness[0.01]}}];
```

> Plot in true scale the same shape with all of its linear dimensions increased by a factor of 5. Use the same PlotRange. Show both on the same axes. How are their area measurements related?

G.9.b.i)   All linear dimensions of a planar region are doubled.

> By what factor does the area measurement change?

G.9.b.ii)   All linear dimensions of a certain solid are doubled.

> By what factor does the volume measurement change?
>
> By what factor does the surface area measurement change?

G.9.c)   A crystal grows in such a way that all the linear dimensions increase by 50%.

> What are the percentages of the increases in the surface area measurement and the volume measurement?

G.9.d.i)   A pyramid whose base is an equilateral triangle 100 feet on a side and whose height is 80 feet is under construction by some really far-out people.

> What will be the finished total volume measurement of this pyramid?

G.9.d.ii)   A pyramid whose base is an equilateral triangle 100 feet on a side and whose finished height is 80 feet is under construction. But due to economic realities, construction is stopped at a height of 57 feet. The resulting structure looks like the bottom part of the pyramid on the left-hand side of the back of a U.S. one dollar bill.

> Measure the volume of this truncated pyramid.
>
> Measure the volume of the unfinished part.

■ G.10) Catfish harvesting

This problem appears only in the electronic version.

LESSON 2.04

Transforming Integrals

Basics

■ **B.1) Breaking more of the code of the integral: Transforming integrals**

B.1.a)

> How do you know that the integrals
>
> $$\int_a^b f'[u[x]]\, u'[x]\, dx \qquad \text{and} \qquad \int_{u[a]}^{u[b]} f'[u]\, du$$
>
> are equal?

Answer: The best way to see why they are equal is to calculate both of them. The second is the easier because the fundamental formula tells you that

$$\int_{u[a]}^{u[b]} f'[u]\, du = f[u]\, \Big[_{u[a]}^{u[b]} = f[u[b]] - f[u[a]].$$

For the first integral, the chain rule tells you that the derivative of $f[u[x]]$ is $f'[u[x]]\, u'[x]$. The fundamental formula steps in to say

$$\int_a^b f'[u[x]]\, u'[x]\, dx = f[u[x]]\, \Big[_a^b = f[u[b]] - f[u[a]].$$

Now you know why

$$\int_a^b f'[u[x]]\, u'[x]\, dx = \int_{u[a]}^{u[b]} f'[u]\, du.$$

Reason: They are both equal to $f[u[b]] - f[u[a]]$.

A point of anxiety might come up: In the integral $\int_{u[a]}^{u[b]} f'[u] \, du$, the lone symbol u is treated as a variable and $u[a]$ and $u[b]$ are numbers. In the integral, $\int_a^b f'[u[x]] \, u'[x] \, dx$, $u[x]$ is a function, x is a variable, and a and b are numbers. This should not cause you any trouble.

B.1.b) Now you know that $\int_a^b f'[u[x]] \, u'[x] \, dx = \int_{u[a]}^{u[b]} f'[u] \, du$.

> What practical use is this?

Answer: Notational magic allows you to take the more complicated integral and replace it with the less complicated but equal integral. Pair them up as follows:

$$f'[u[x]] \longleftrightarrow f'[u]$$
$$u'[x] \, dx \longleftrightarrow du$$
$$\int_a^b \longleftrightarrow \int_{u[a]}^{u[b]} \ .$$

Lots of folks like to call this a "transformation" of a hard integral into an easy integral.

B.1.c) Here is *Mathematica*'s calculation of $\int_0^\pi \cos[x^2] \, 2 \, x \, dx$:

In[1]:=
```
Clear[x]
MathematicaCalculation = Integrate[Cos[x^2] 2 x,{x,0,Pi}]
```
Out[1]=
```
      2
Sin[Pi ]
```

> Use a transformation to explain where this result comes from.

Answer: The integral at the center stage is $\int_0^\pi \cos[x^2] \, 2 \, x \, dx$. The key is that $2 \, x$ is the derivative of x^2, so go with

$$u[x] = x^2.$$

This gives the pairings

$$\cos[x^2] \longleftrightarrow \cos[u]$$
$$2 \, x \, dx = u'[x] \, dx \longleftrightarrow du$$
$$\int_0^\pi = \int_a^b \longleftrightarrow \int_{u[a]}^{u[b]} = \int_0^{\pi^2} \ .$$

So, $\int_0^\pi \cos[x^2]\, 2\, x\, dx = \int_0^{\pi^2} \cos[u]\, du$. In other words, the substitution $u[x] = x^2$ transforms $\int_0^\pi \cos[x^2]\, 2\, x\, dx$ into $\int_0^{\pi^2} \cos[u]\, du$.

The second integral is cake. You look for a function $f[u]$ with $f'[u] = \cos[u]$. Here's one:

In[2]:=
```
Clear[f,u]; f[u_] = Sin[u];
```

Check:

In[3]:=
```
f'[u]
```
Out[3]=
```
Cos[u]
```

And now you can say with confidence and authority that

$$\int_0^\pi \cos[x^2]\, 2\, x\, dx = \int_0^{\pi^2} \cos[u]\, du$$

is given by:

In[4]:=
```
f[Pi^2] - f[0^2]
```
Out[4]=
```
     2
Sin[Pi ]
```

Check with *Mathematica*'s calculation.

In[5]:=
```
MathematicaCalculation
```
Out[5]=
```
     2
Sin[Pi ]
```

Got it.

B.1.d) Here is *Mathematica*'s calculation of $\int_0^\pi e^{\sin[a x]} \cos[a\, x]\, dx$:

In[6]:=
```
Clear[x,a]
MathematicaCalculation = Integrate[(E^(Sin[a x])) Cos[a x],{x,0,Pi}]
```
Out[6]=
```
          Sin[a Pi]
     1   E
  -(-) + ---------
     a       a
```

Use a transformation to explain where this result comes from.

Answer: The spotlight is on

$$\int_0^\pi e^{\sin[a x]} \cos[a\,x]\,dx.$$

The function $a \cos[a\,x]$ is the derivative of $\sin[a\,x]$; so go with

$$u[x] = \sin[a\,x].$$

The "$a \cos[a\,x]$" term you need is not immediately available; react by rewriting the integral as

$$\left(\frac{1}{a}\right) \int_0^\pi e^{\sin[a x]}\, a\,\cos[a\,x]\,dx.$$

This is a legal step because a is a constant.

Now there is the $a \cos[a\,x]$ term right where you want it. This gives the pairings

$$e^{\sin[a x]} \longleftarrow\!\!\!\!\longrightarrow e^u$$

$$a\,\cos[a\,x]\,dx = u'[x]\,dx \longleftarrow\!\!\!\!\longrightarrow du$$

$$\left(\frac{1}{a}\right) \int_0^\pi \longleftarrow\!\!\!\!\longrightarrow \left(\frac{1}{a}\right) \int_0^{\sin[a\pi]}.$$

So

$$\int_0^\pi e^{\sin[a x]} \cos[a\,x]\,dx = \left(\frac{1}{a}\right) \int_0^\pi e^{\sin[a x]}\, a\,\cos[a\,x]\,dx$$

$$= \left(\frac{1}{a}\right) \int_0^{\sin[a\pi]} e^u\,du.$$

Now look for a function $f[u]$ whose derivative is e^u. One such is:

In[7]:=
```
Clear[f,u]; f[u_] = E^u;
```

Now you know for sure that

$$\int_0^\pi e^{\sin[a x]} \cos[a\,x]\,dx = \left(\frac{1}{a}\right) \int_0^{\sin[a\pi]} e^u\,du$$

is given by:

In[8]:=
```
(1/a)(f[Sin[a Pi]] - f[0])
```

Out[8]=

$$\frac{-1 + E^{\text{Sin}[a\ \text{Pi}]}}{a}$$

Compare:

In[9]:=
```
Together[MathematicaCalculation]
```

Out[9]=

$$\frac{\text{Sin[a Pi]}}{-1 + \text{E}}$$
$$a$$

Got it.

B.1.e)   Here is *Mathematica*'s calculation of $\int_0^1 \sqrt{1 - x^2}\ dx$:

In[10]:=
```
Clear[x]
MathematicaCalculation = Integrate[Sqrt[1 - x^2],{x,0,1}]
```

Out[10]=

$$\frac{\text{Pi}}{4}$$

> Use a transformation to explain where this result comes from.

Answer: The integral under study is $\int_0^1 \sqrt{1 - x^2}\ dx$. Make the wild card substitution

$$x = \sin[t].$$

This gives you the pairings

$$\sqrt{1 - x^2} \longleftrightarrow \sqrt{1 - \sin[t]^2} = \cos[t];$$
$$dx \longleftrightarrow dt;$$
$$\int_0^1 \longleftrightarrow \int_0^{\pi/2}$$

because $\sin[0] = 0$ and $\sin[\pi/2] = 1$. Now you know that

$$\int_0^1 \sqrt{1 - x^2}\ dx = \int_0^{\pi/2} \cos[t]\ \cos[t]\ dt = \int_0^{\pi/2} \cos[t]^2\ dt.$$

To calculate $\int_0^{\pi/2} \cos[t]^2\ dt$, apply a trig identity to $\cos[t]^2$:

In[11]:=
```
Clear[t]; Expand[Cos[t]^2,Trig->True]
```

Out[11]=

$$\frac{1}{2} + \frac{\text{Cos[2 t]}}{2}$$

Now you know that

$$\int_0^1 \sqrt{1 - x^2}\ dx = \int_0^{\pi/2} \cos[t]^2\ dt = \int_0^{\pi/2} \left(\frac{1}{2} + \frac{\cos[2\,t]}{2} \right)\ dt.$$

To complete the calculation, look for a function $f[t]$ with $f'[t] = 1/2 + \cos[2\,t]/2$. Here is one such:

In[12]:=
```
Clear[f,t]; f[t_] = t/2 + Sin[2 t]/4;
```

Check whether $f'[t] = 1/2 + \cos[2\,t]/2$:

In[13]:=
```
f'[t]
```
Out[13]=

$$\frac{1}{2} + \frac{\text{Cos}[2\ t]}{2}$$

Good; now you can say with confidence and considerable authority that

$$\int_0^1 \sqrt{1 - x^2}\ dx = \int_0^{\pi/2} \left(\frac{1}{2} + \frac{\cos[2\,t]}{2} \right) dt$$

is given by:

In[14]:=
```
f[Pi/2] - f[0]
```
Out[14]=

$$\frac{\text{Pi}}{4}$$

Compare:

In[15]:=
```
MathematicaCalculation
```
Out[15]=

$$\frac{\text{Pi}}{4}$$

Got it.

B.1.f) Look at *Mathematica*'s calculation of

$$\int_3^5 e^{-((x-3)/2)^2}\ dx\ :$$

In[16]:=
```
Clear[x]
Integrate[(E^(-((x - 3)/2)^2)),{x,3,5}]
```
Out[16]=
```
Sqrt[Pi] Erf[1]
```

Remembering that Erf[x] is defined by

$$\text{Erf}[x] = \left(\frac{2}{\sqrt{\pi}} \right) \int_0^x e^{-t^2}\ dt,$$

you see that *Mathematica* is telling you that $\int_3^5 e^{-((x-3)/2)^2} \, dx$ transforms into

$$\sqrt{\pi} \left(\frac{2}{\sqrt{\pi}} \right) \int_0^1 e^{-t^2} \, dt = 2 \int_0^1 e^{-t^2} \, dt.$$

Use a substitution to explain why

$$\int_3^5 e^{-((x-3)/2)^2} \, dx = 2 \int_0^1 e^{-t^2} \, dt.$$

Answer: To do this, take $\int_3^5 e^{-((x-3)/2)^2} \, dx$ and make the substitution

$$t = \frac{x-3}{2}.$$

This gives $dt = (1/2) \, dx$, which is the same as $2 \, dt = dx$. This gives the pairings

$$e^{-((x-3)/2)^2} \longleftrightarrow e^{-t^2};$$

$$dx \longleftrightarrow 2 \, dt;$$

$$\int_3^5 \longleftrightarrow \int_{(3-3)/2}^{(5-3)/2} = \int_0^1.$$

The upshot is

$$\int_3^5 e^{-((x-3)/2)^2} \, dx = \int_0^1 e^{-t^2} \, 2 \, dt = 2 \int_0^1 e^{-t^2} \, dt.$$

Done.

■ B.2) Measuring area under curves given parametrically

B.2.a) Here's a curve specified by parametric formulas:

```
In[17]:=
  tlow = 0; thigh = 4 Pi; Clear[x,y,t]
  {x[t_],y[t_]} = {t + Sin[t/2],2 + Cos[t]};
  curveplot = ParametricPlot[
  {x[t],y[t]},{t,tlow,thigh},
  PlotStyle->{{Red,Thickness[0.01]}},
  PlotRange->{0,3},AxesLabel->{"x","y"}];
```

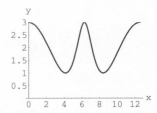

Measuring the area under this curve and over the x-axis is a snap. You just calculate

$$\int_{tlow}^{thigh} y[t] \, x'[t] \, dt :$$

In[18]:=
```
Integrate[y[t] x'[t],{t,tlow,thigh}]
```
Out[18]=
```
8 Pi
```

> Use a transformation to explain why this integral actually measures the area under this curve and over the x-axis.

Answer: Take another look at the curve. If you can come up with a formula for the function $f[x]$ such that the curve above is the plot of $f'[x]$, then you can confirm the calculation by calculating:

In[19]:=
```
{xlow,xhigh} = {x[tlow],x[thigh]}
```
Out[19]=
```
{0, 4 Pi}
```

and then using the fundamental formula to measure the area by writing

$$\int_{\text{xlow}}^{\text{xhigh}} f'[x]\,dx = f[\text{xhigh}] - f[\text{xlow}].$$

Theoretically, this is fine, but to use it you have to come up with a formula for $f[x]$. Part of the art of mathematics is avoiding needless calculations and formulas. It turns out that you don't need the exact formula for $f[x]$ to see why the integral

$$\int_{\text{tlow}}^{\text{thigh}} y[t]\,x'[t]\,dt$$

actually makes the measurement. Reason: You can transform

$$\int_{\text{xlow}}^{\text{xhigh}} f'[x]\,dx$$

into

$$\int_{\text{tlow}}^{\text{thigh}} y[t]\,x'[t]\,dt.$$

Here's how it goes: Without having your hands on a formula for $f[x]$, look at $\int_{\text{xlow}}^{\text{xhigh}} f'[x]\,dx$. Make the substitution

$$x = x[t].$$

This gives the pairings:

$$f'[x] \longleftrightarrow f'[x[t]]$$
$$dx \longleftrightarrow x'[t]\,dt$$
$$\int_{\text{xlow}}^{\text{xhigh}} \longleftrightarrow \int_{\text{tlow}}^{\text{thigh}}.$$

So,

$$\int_{\text{xlow}}^{\text{xhigh}} f'[x]\, dx = \int_{\text{tlow}}^{\text{thigh}} f'[x[t]]\, x'[t]\, dt.$$

But wait a minute! You know

$$f'[x[t]] = y[t],$$

and this gives you the formula you want:

$$\int_{\text{xlow}}^{\text{xhigh}} f'[x]\, dx = \int_{\text{tlow}}^{\text{thigh}} f'[x[t]]\, x'[t]\, dt = \int_{\text{tlow}}^{\text{thigh}} y[t]\, x'[t]\, dt.$$

Done! The theoretical existence of $f[x]$ was all that was needed to explain the formula. This is another example of a situation in which theory helped to avoid a miserable calculation.

B.2.b) Here is one arch of the cycloid

$$x = x[t] = 10\, t - 10\, \sin[t]$$

and

$$y = y[t] = 10 - 10\, \cos[t].$$

In[20]:=
```
tlow = 0; thigh = 2 Pi; Clear[x,y,t];
x[t_] = 10 t - 10 Sin[t]; y[t_] = 10 - 10 Cos[t];
arch = ParametricPlot[{x[t],y[t]},{t,0,2 Pi},
AspectRatio->Automatic,PlotStyle->
{{Blue,Thickness[0.01]}},AxesLabel->{"x","y"}];
```

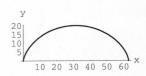

> Measure the area under the arch and over the x-axis.

Answer: Here it comes:

In[21]:=
```
Integrate[y[t] x'[t],{t,tlow,thigh}]
```

Out[21]=
```
300 Pi
```

Done.

■ B.3) Bell-shaped curves and Gauss's normal law

Calculus&*Mathematica* is pleased to acknowledge that this problem leans heavily on the material in Chapter 3 of Mark Kac's autobiography, *Enigmas of Chance*, University of California Press, Berkeley, 1985.

Paraphrasing Mark Kac: Of all curves, the most constantly encountered and the only one claimed as its own by the mathematical, physical, biological, and social sciences is the famous bell-shaped curve

$$\text{bell}[x] = \left(\frac{1}{\sqrt{2\,\pi}}\right) e^{-x^2/2}.$$

Here's how this beauty looks:

In[22]:=
```
Clear[x,bell]
bell[x_] = (1/Sqrt[2 Pi]) E^((-x^2)/2);
Plot[bell[x],{x,-5,5},
PlotStyle->{{Blue,Thickness[0.01]}},
PlotRange->All,AxesLabel->{"x","bell[x]"},
Epilog->Text["Basic bell-shaped curve",
{-E,1/E}]];
```

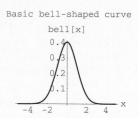

Even its formula,

$$\text{bell}[x] = \left(\frac{1}{\sqrt{2\,\pi}}\right) e^{-x^2/2},$$

unites two of the most alluring constants in all of science, e and π. Here are two more seductive features of bell$[x]$: $\int_{-\infty}^{\infty}\text{bell}[x]\,dx = 1$:

In[23]:=
```
Integrate[bell[x],{x,-Infinity,Infinity}]
```

Out[23]=
```
1
```

Also, bell$[x] = D[\text{Erf}[x/\sqrt{2}\,]/2, x]$

In[24]:=
```
D[Erf[x/Sqrt[2]]/2,x] == bell[x]
```

Out[24]=
```
True
```

B.3.a.i) | How does the fact that
$$\text{bell}[x] = D\left[\frac{\text{Erf}[x/\sqrt{2}\,]}{2}, x\right]$$
begin to explain why
$$\int_{-\infty}^{\infty}\text{bell}[x]\,dx = 1?$$

Answer: Look at a plot of Erf$[x/\sqrt{2}\,]/2$:

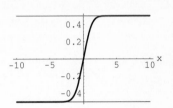

```
In[25]:=
  Clear[x]
  Plot[{1/2,-1/2,
  Erf[x/Sqrt[2]]/2},{x,-10,10},
  PlotStyle->{{Red},{Red},
  {Blue,Thickness[0.01]}},
  AxesLabel->{"x",""}];
```

As the plot shows,

$$\lim_{x \to \infty} \frac{\mathrm{Erf}[x/\sqrt{2}\,]}{2} = \frac{1}{2},$$

$$\lim_{x \to -\infty} \frac{\mathrm{Erf}[x/\sqrt{2}\,]}{2} = -\frac{1}{2}$$

and

$$\frac{\mathrm{Erf}[0/\sqrt{2}\,]}{2} = \frac{\mathrm{Erf}[0]}{2} = 0.$$

Now because bell[x] is the derivative of $\mathrm{Erf}[x/\sqrt{2}\,]/2$, the fundamental formula tells you that

$$\int_0^\infty \mathrm{bell}[x]\,dx = \lim_{t \to \infty} \int_0^t \mathrm{bell}[x]\,dx$$

$$= \lim_{t \to \infty} \frac{\mathrm{Erf}[t/\sqrt{2}\,]}{2} - \frac{\mathrm{Erf}[0]}{2}$$

$$= \frac{1}{2} - 0 = \frac{1}{2}.$$

Similarly,

$$\int_{-\infty}^0 \mathrm{bell}[x]\,dx = \lim_{t \to -\infty} \int_t^0 \mathrm{bell}[x]\,dx$$

$$= \frac{\mathrm{Erf}[0]}{2} - \lim_{t \to -\infty} \frac{\mathrm{Erf}[t/\sqrt{2}\,]}{2}$$

$$= 0 - \left(-\frac{1}{2}\right) = \frac{1}{2}.$$

Accordingly,

$$\int_{-\infty}^\infty \mathrm{bell}[x]\,dx = \int_{-\infty}^0 \mathrm{bell}[x]\,dx + \int_0^\infty \mathrm{bell}[x]\,dx = \frac{1}{2} + \frac{1}{2} = 1.$$

B.3.a.ii) Put

$$\mathrm{normal}[x, \mathrm{mean}, \mathrm{dev}] = \frac{\mathrm{bell}[(x - \mathrm{mean})/\mathrm{dev}]}{\mathrm{dev}}:$$

```
In[26]:=
  Clear[x,bell]
  bell[x_] = (1/Sqrt[2 Pi]) E^((-x^2)/2);
  Clear[normal,mean,dev]
  normal[x_,mean_,dev_] = bell[(x - mean)/dev]/dev;
```

Here's a plot of normal$[x, 2, 3]$ as a function of x:

In[27]:=
```
mean = 2; dev = 3;
dome = Plot[normal[x,mean,dev],
{x,mean - 10,mean + 10},
PlotStyle->{{Blue,Thickness[0.01]}},
PlotRange->All,AxesOrigin->{mean,0}];
```

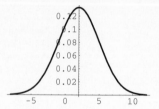

Here's a plot of normal$[x, 2, 0.5]$ as a function of x:

In[28]:=
```
mean = 2; dev = 0.5;
spike = Plot[normal[x,mean,dev],
{x,mean - 10,mean + 10},
PlotStyle->{{Red,Thickness[0.01]}},
PlotRange->All,AxesOrigin->{mean,0}];
```

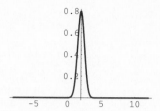

Here they are together:

In[29]:=
```
Show[dome,spike];
```

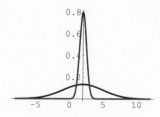

Observations:

→ The plot of normal$[x, \text{mean}, \text{dev}]$ is a modification of the basic bell-shaped curve.

→ The plot of normal$[x, \text{mean}, \text{dev}]$ crests when $x = \text{mean}$.

→ Smaller values of dev force a thinner, higher bell than do larger values of dev.

Try some more to anchor this in your mind.

After you get the idea, go with cleared values of mean and dev and explain why

$$\int_{-\infty}^{\infty} \text{normal}[x, \text{mean}, \text{dev}] \, dx = 1 \qquad \text{(when dev}> 0)$$

by transforming the integral

$$\int_{-\infty}^{\infty} \text{normal}[x, \text{mean}, \text{dev}] \, dx$$

into the integral $\int_{-\infty}^{\infty} \text{bell}[x] \, dx = 1$.

Answer: Remember,

$$\text{normal}[x, \text{mean}, \text{dev}] = \frac{\text{bell}[(x - \text{mean})/\text{dev}]}{\text{dev}}.$$

This tells you that

$$\int_{-\infty}^{\infty} \text{normal}[x, \text{mean}, \text{dev}] \, dx = \int_{-\infty}^{\infty} \text{bell}\left[\frac{x - \text{mean}}{\text{dev}}\right] \left(\frac{1}{\text{dev}}\right) dx.$$

Now make the substitution

$$t = \frac{x - \text{mean}}{\text{dev}}.$$

This gives the pairings

$$\text{bell}\left[\frac{x - \text{mean}}{\text{dev}}\right] \longleftrightarrow \text{bell}[t]$$

$$\frac{1}{\text{dev}} \, dx = \longleftrightarrow dt$$

$$\int_{-\infty}^{\infty} \longleftrightarrow \int_{-\infty}^{\infty}$$

because dev > 0 and $t = -\infty$ when $x = -\infty$ and $t = \infty$ when $x = \infty$. So,

$$\int_{-\infty}^{\infty} \text{normal}[x, \text{mean}, \text{dev}] \, dx = \int_{-\infty}^{\infty} \text{bell}\left[\frac{x - \text{mean}}{\text{dev}}\right] \left(\frac{1}{\text{dev}}\right) dx$$

$$= \int_{-\infty}^{\infty} \text{bell}[t] \, dt$$

$$= \int_{-\infty}^{\infty} \text{bell}[x] \, dx = 1$$

by what you know from part B.3.a.i) above. This explains why

$$\int_{-\infty}^{\infty} \text{normal}[x, \text{mean}, \text{dev}] \, dx = 1$$

provided dev > 0.

B.3.b) Still paraphrasing Mark Kac: All observations, constructions, and measurements are subject to error. You can classify errors or deviations as follows:

→ Gross errors or deviations,

→ Systematic errors or deviations,

→ Random errors or deviations.

Sufficient care can eliminate gross errors, gross deviations, systematic errors, and systematic deviations. But random errors and random deviations are another story. The great German mathematician Karl Friedrich Gauss (1777–1855) designed a theory to deal with random errors or deviations. Here is Gauss's theory in a nutshell:

Start with a measurement, construction, or observation M. Call the true value of the measurement, construction, or observation by the name "mean." Gauss's theory says that often you can find a value dev, called "standard deviation," so that the fraction of the time that M actually measures out so that $a \leq M \leq b$ is given by

$$\int_a^b \text{normal}[x, \text{mean}, \text{dev}] \, dx.$$

For instance, if

mean $= 70$

and

dev $= 10$,

then the fraction of the time that M actually measures out so that $65 \leq M \leq 75$ is given by

$$\int_{65}^{75} \text{normal}[x, 70, 10] \, dx :$$

In[30]:=
```
Clear[x,bell]
bell[x_] = (1/Sqrt[2 Pi]) E^((-x^2)/2);
Clear[normal,mean,dev]
normal[x_,mean_,dev_] = bell[(x - mean)/dev]/dev;
NIntegrate[normal[x,70,10],{x,65,75}]
```

Out[30]=
```
0.382925
```

About 38% of the time, M comes out between 65 and 75. To see what fraction of the time M comes out between 60 and 80, calculate

$$\int_{60}^{80} \text{normal}[x, 70, 10] \, dx :$$

In[31]:=
```
NIntegrate[normal[x,70,10],{x,60,80}]
```

Out[31]=
```
0.682689
```

About 68% of the time, M comes out between 60 and 80. But if you can cut dev from 10 to 5, then the fraction of the time that M comes out between 60 and 80 is given by

$$\int_{60}^{80} \text{normal}[x, 70, 5] \, dx :$$

In[32]:=
```
NIntegrate[normal[x,70,5],{x,60,80}]
```

Out[32]=
```
0.9545
```

Now with dev reduced to dev = 5, M comes out between 60 and 80 about 95% of the time.

> Use a plot to explain why it turns out that with the lower value of dev, M came out between 60 and 80 more of the time than it did with the higher value of dev.

Answer: Here are plots of normal$[x, \text{mean}, 5]$ and normal$[x, \text{mean}, 10]$ with mean = 70 in both:

In[33]:=
```
a = 60; b = 80; mean = 70;
Plot[{normal[x,mean,10],normal[x,mean,5]},
{x,mean -20,mean + 20},
PlotStyle->{{Blue,Thickness[0.02]},
{Red,Thickness[0.01]}},AxesOrigin->{mean,0},
Epilog->{Line[{{a,0},{a,normal[a,mean,10]}}],
Line[{{b,0},{b,normal[b,mean,10]}}]}];
```

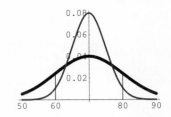

The plot of normal$[x, 70, 10]$ is thicker than the plot of normal$[x, 70, 5]$.

As x goes from $-\infty$ to ∞, the area between the x-axis and each curve is 1 because

$$\int_{-\infty}^{\infty} \text{normal}[x, \text{mean}, \text{dev}] \, dx = 1$$

no matter what mean and dev are as long as dev > 0. The plot of normal$[x, 70, 5]$ (spiked plot) packs more of this area between $x = 60$ and $x = 80$ than does the plot of normal$[x, 70, 10]$ (rounded plot). That's why

$$\int_{60}^{80} \text{normal}[x, 70, 5] \, dx > \int_{60}^{80} \text{normal}[x, 70, 10] \, dx.$$

Moral: The lower the standard deviation (dev), the more likely M will come out near its mean.

B.3.c.i) Here are the results of a study of a sample of Canadian nine-year-old boys in the form $\{x, y\}$ with

> $x =$ weight within 2.5 pounds

and

> $y =$ number with weight x.

All weights are given within an accuracy of 2.5 pounds.

In[34]:=
```
data = {{42, 20},{47,146},{52,553},{57,979},{62,1084},{67,807},{72,468},
{77,180},{82,103},{87,41},{92,25},{97,21},{102,11},{107,5},{112,5},{117,3}};
```

Calculate the average weight:

In[35]:=
```
Clear[k]
numberofboys = Sum[data[[k,2]],{k,1,Length[data]}];
N[Sum[data[[k,1]] data[[k,2]],{k,1,Length[data]}]/numberofboys]
```
Out[35]=
```
62.9099
```

The average weight of a boy in the sample is about 63 pounds:

In[36]:=
```
average = 63;
```

Take a look at a plot of the data:

In[37]:=
```
dataplot = ListPlot[data,
PlotStyle->{Red,PointSize[0.03]},
AxesLabel->{"weight",
"number with\n this weight"},
AxesOrigin->{average,0}];
```

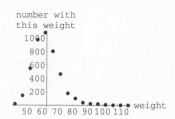

Source: Mark Kac, *Enigmas of Chance*, University of California Press, Berkeley, 1985.

Is that bell-shaped or what?

Come up with a value of mean, a value of t, and a value of dev so that

$$t \, \text{normal}[x, \text{mean}, \text{dev}]$$

gives a good fit of these data.

Answer: Remember, the average weight is 63 pounds.

Start by finding t and dev so that the plot of

$$t \, \text{normal}[x, 63, \text{dev}]$$

does a good job of modeling the data. Make refinements later. Look at the data:

In[38]:=
```
data
```
Out[38]=
```
{{42, 20},{47, 146},{52, 553},{57, 979},{62, 1084},
{67, 807},{72, 468},{77, 180},{82, 103},{87, 41},
{92, 25},{97, 21},{102, 11},{107, 5},{112, 5},{117, 3}}
```

The biggest is 1084; that's where you want the top of the bell to be. Notice that this happens near $x = 63$, the average weight:

In[39]:=
```
Show[dataplot];
```

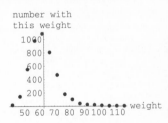

This means that it's a good idea at first to go with t and dev so that

$$1084 = t \, \text{normal}[63, 63, \text{dev}].$$

In[40]:=
```
Clear[x,bell,normal,mean,dev,t]
bell[x_] = (1/Sqrt[2 Pi]) E^((-x^2)/2);
normal[x_,mean_,dev_] = bell[(x - mean)/dev]/dev;
1084 == t normal[63,63,dev]
```

Out[40]=

$$1084 == \frac{t}{\text{dev Sqrt}[2 \text{ Pi}]}$$

Read off:

In[41]:=
```
Clear[t,dev];
t[dev_] = 1084 dev Sqrt[2 Pi]
```

Out[41]=
```
1084 dev Sqrt[2 Pi]
```

Now try dev = 4:

In[42]:=
```
dev = 4;
fitplot = Plot[t[dev] normal[x,63,dev],
{x,0,120},PlotStyle->{{Blue,Thickness[0.01]}},
DisplayFunction->Identity];
Show[dataplot,fitplot,PlotRange->All,
AxesOrigin->{average,0},
DisplayFunction->$DisplayFunction];
```

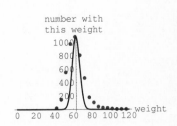

Increase dev to make the bell fatter:

In[43]:=
```
dev = 8;
fitplot = Plot[t[dev] normal[x,63,dev],
{x,0,120},PlotStyle->{{Blue,Thickness[0.01]}},
DisplayFunction->Identity];
Show[dataplot,fitplot,
PlotRange->All,AxesOrigin->{average,0},
DisplayFunction->$DisplayFunction];
```

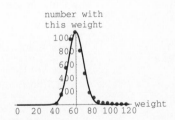

Lookin' pretty good. Refine it by shifting the plot a wee bit to the left:

```
In[44]:=
  dev = 8;
  fitplot = Plot[t[dev] normal[x,62,dev],
  {x,0,120},PlotStyle->{{Blue,Thickness[0.01]}},
  DisplayFunction->Identity];
  Show[dataplot,fitplot,PlotRange->All,
  AxesOrigin->{62,0},
  DisplayFunction->$DisplayFunction];
```

That's really a pretty good fit. You might try for a better fit, but the data are rough enough that a super-good fit isn't realistic.

B.3.c.ii) Look again at the data:

```
In[45]:=
  data
```

```
Out[45]=
  {{42, 20},{47, 146},{52, 553},{57, 979},{62, 1084},
  {67, 807},{72, 468},{77, 180},{82, 103},{87, 41},
  {92, 25},{97, 21},{102, 11},{107, 5},{112, 5},{117, 3}}
```

This is in the form $\{x, y\}$ with

x = weight within 2.5 pounds

and

y = number with weight x.

Convert to a new form $\{x, y\}$ so that

y = the fraction of the total sample with weight x :

```
In[46]:=
  Clear[k]
  fractions = Table[{data[[k,1]],
  N[data[[k,2]]/numberofboys]},
  {k,1,Length[data]}]
```

```
Out[46]=
  {{42, 0.00449337}, {47, 0.0328016}, {52, 0.124242},
  {57, 0.219951}, {62, 0.243541}, {67, 0.181308},
  {72, 0.105145}, {77, 0.0404404}, {82, 0.0231409},
  {87, 0.00921141}, {92, 0.00561672}, {97, 0.00471804},
  {102, 0.00247135}, {107, 0.00112334},
  {112, 0.00112334}, {117, 0.000674006}}
```

Remember, the weights are given to the nearest 2.5 pounds.

Here are the fractions for the boys whose weights measure out between 49.5 pounds and 79.5 pounds:

```
In[47]:=
  Table[{fractions[[k,1]],fractions[[k,2]]},{k,3,8}]
```

Out[47]=
```
{{52, 0.124242}, {57, 0.219951}, {62, 0.243541},
 {67, 0.181308}, {72, 0.105145}, {77, 0.0404404}}
```

The fraction of the boys whose weights measure out between 49.5 pounds and 79.5 pounds is:

In[48]:=
```
Sum[fractions[[k,2]],{k,3,8}]
```

Out[48]=
```
0.914626
```

Now take the function you got from the fit in part B.3.c.i), normal$[x, 62, 8]$, and look at

$$\int_{49.5}^{79.5} \text{normal}[x, 62, 8] \, dx :$$

In[49]:=
```
NIntegrate[normal[x,62,8],{x,49.5,79.5}]
```

Out[49]=
```
0.926562
```

This is very close to the fraction of the boys whose weights measure out between 49.5 pounds and 79.5 pounds. Try it again.

In[50]:=
```
fractions
```

Out[50]=
```
{{42, 0.00449337}, {47, 0.0328016}, {52, 0.124242},
 {57, 0.219951}, {62, 0.243541}, {67, 0.181308},
 {72, 0.105145}, {77, 0.0404404}, {82, 0.0231409},
 {87, 0.00921141}, {92, 0.00561672}, {97, 0.00471804},
 {102, 0.00247135}, {107, 0.00112334},
 {112, 0.00112334}, {117, 0.000674006}}
```

Remember, the weights are given to the nearest 2.5 pounds.

Here are the fractions for the boys whose weights measure out between 54.5 pounds and 69.5 pounds:

In[51]:=
```
Table[{fractions[[k,1]],fractions[[k,2]]},{k,4,6}]
```

Out[51]=
```
{{57, 0.219951}, {62, 0.243541}, {67, 0.181308}}
```

The fraction of the boys whose weights measure out between 54.5 pounds and 69.5 pounds is:

In[52]:=
```
Sum[fractions[[k,2]],{k,4,6}]
```

Out[52]=
```
0.644799
```

Now take the function you got from the fit in part B.3.c.i), normal$[x, 62, 8]$, and look at

$$\int_{54.5}^{69.5} \text{normal}[x, 62, 8] \, dx :$$

In[53]:=
```
NIntegrate[normal[x,62,8],{x,54.5,69.5}]
```

Out[53]=
```
0.651499
```

This is very close to the fraction of the boys whose weights measure out between 54.5 pounds and 69.5 pounds.

What's going on here?

Answer: You're seeing Gauss's normal law in action. Once you have the good fit of the data as done above, you can estimate the fraction of weights that came out between a and b by calculating $\int_a^b \text{normal}[x, 62, 8] \, dx$. To see why this works, look at the last plot:

In[54]:=
```
Show[dataplot,fitplot,
PlotRange->All,AxesOrigin->{62,0},
DisplayFunction->$DisplayFunction];
```

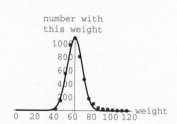

The curve plotted here is a multiple of normal$[x, 62, 8]$. Here's the mathematical beauty of it: Because

$$\int_{-\infty}^{\infty} \text{normal}[x, \text{mean}, \text{dev}] \, dx = 1$$

no matter what mean and dev are, you are guaranteed that

$$\int_{-\infty}^{\infty} \text{normal}[x, 62, 8] \, dx = 1.$$

The upshot:

$$\int_a^b \text{normal}[x, 62, 8] \, dx$$

measures the fraction of the area under the plotted curve over the interval $[a, b]$ on the x-axis. And because this plotted curve is a good approximation of the data, $\int_a^b \text{normal}[x, 62, 8] \, dx$ also estimates the fraction of the weights that came out between a and b. That's all there is to it. This is especially useful when you want to estimate fractions of the weights that aren't specifically detailed in the original data. For instance, you can estimate the fraction of the boys whose weights are between 50 and 80 pounds by calculating: $\int_{50}^{80}$ normal$[x, 62, 8] \, dx$:

In[55]:=
```
NIntegrate[normal[x,62,8],{x,50,80}]
```
Out[55]=
```
0.920968
```

About 93% of the boys had weights between 50 and 80 pounds. In other words, if you select one of the kids at random, about 93% of the time you will select a kid whose weight is between 50 and 80 pounds. The fancy folks say that a kid has weight between 50 and 80 pounds with probability 0.93. To see what fraction of the kids have weights over 80 pounds, calculate $\int_{80}^{\infty}$ normal$[x, 62, 8] \, dx$:

In[56]:=
```
NIntegrate[normal[x,62,8],{x,80,Infinity},AccuracyGoal->2]
```
Out[56]=
```
0.0122686
```

About 2% of the boys had weights over 80 pounds. In other words, if you select one of the kids at random, about 2% of the time you will select a kid whose weight is over 80 pounds.

B.3.d.i) | What do folks mean when they say that a measurement M is normally distributed with
$$\text{mean} = m$$
and
$$\text{standard deviation} = s?$$

Answer: They mean that the fraction of the time that the measurement comes out between a and b is given by

$$\int_a^b \text{normal}[x, m, s] \, dx.$$

The work above indicates that the weights of the Canadian boys are approximately normally distributed with mean = 62 and standard deviation = 8 because you can get a reasonable estimate of the fraction of the weights that come out between a and b by calculating $\int_a^b$ normal$[x, 62, 8] \, dx$.

B.3.d.ii) | When you suspect that a measurement M is normally distributed, how do you try to determine the specific values of mean = m and standard deviation = s?

Answer: You get good data and then try to fit the data as done earlier. Once you have a reasonable fit, you can read off the mean $= m$ and standard deviation $= s$ as done above.

B.3.d.iii) Are all measurements M normally distributed?

Answer: No. Statistics courses look at other possibilities. You can get a good idea of whether a given measurement is normally distributed by collecting data and trying to fit them as you did above. If you are successful, then you know that the measurement M is normally distributed and you can read off the mean and standard deviation. If such a fit is impossible, then you decide that the measurement is not normally distributed.

Tutorials

■ T.1) Transforming integrals

This problem appears only in the electronic version.

■ T.2) Transforming integrals to help understand *Mathematica* output

This problem appears only in the electronic version.

■ T.3) Measuring area inside closed curves

T.3.a.i) Here's a plot of the parametric curve
$$x = x[t] = 5 \cos[t]$$
and
$$y = y[t] = 4 \sin[t]^3.$$

In[1]:=
```
Clear[t,x,y]; tlow = 0; thigh = 2 Pi;
x[t_] = 5 Cos[t]; y[t_] = 4 Sin[t]^3;
redlips = ParametricPlot[
{x[t],y[t]},{t,tlow,thigh},
AspectRatio->Automatic,
PlotStyle->{{Red,Thickness[0.01]}}];
```

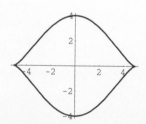

Most folks call this kind of curve by the name "closed curve." It starts at:

In[2]:=
```
{x[tlow],y[tlow]}
```

Out[2]=
```
{5, 0}
```

And it ends at:

In[3]:=
```
{x[thigh],y[thigh]}
```

Out[3]=
```
{5, 0}
```

It starts right where it stops.

Measure the area enclosed by this curve.

Answer: Take another look. Evidently the area the curve encloses is four times the area under the curve and over $[0, 5]$ on the x-axis. Let's see which t's give the plot in the first quadrant: It looks like $0 \le t \le \pi/2$ will do the trick, but check with a plot:

In[4]:=
```
ParametricPlot[
{x[t],y[t]},{t,0,Pi/2},
AspectRatio->Automatic,
PlotStyle->{{Red,Thickness[0.01]}}];
```

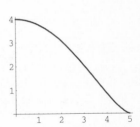

Good. Now check to see which t's correspond to $x = 0$ and $x = 5$:

In[5]:=
```
x[0]
```

Out[5]=
```
5
```

In[6]:=
```
x[Pi/2]
```

Out[6]=
```
0
```

The measurement you want is $4 \int_0^5 y \, dx = 4 \int_{\pi/2}^0 y[t] \, x'[t] \, dt$ square units:

In[7]:=
```
areameasure = 4 NIntegrate[y[t] x'[t],{t,Pi/2,0}]
```

Out[7]=
```
47.1239
```

Looks good!

T.3.a.ii) Here's the same curve as in part T.3.a.i):

In[8]:=
```
Clear[t,x,y]; tlow = 0; thigh = 2 Pi;
x[t_] = 5 Cos[t]; y[t_] = 4 Sin[t]^3;
redlips = ParametricPlot[
{x[t],y[t]},{t,tlow,thigh},
AspectRatio->Automatic,
PlotStyle->{{Red,Thickness[0.01]}}];
```

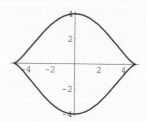

In part T.3.a.i), you found that the area enclosed by this curve measures out to:

In[9]:=
```
areameasure = 4 NIntegrate[y[t] x'[t],{t,Pi/2,0}]
```

Out[9]=
```
47.1239
```

See what happens when you calculate $\int_{\text{tlow}}^{\text{thigh}} y[t]\, x'[t]\, dt$:

In[10]:=
```
Integrate[y[t] x'[t],{t,tlow,thigh}]
```

Out[10]=
```
-15 Pi
```

This gives you the negative of the area measurement, and this cannot be an accident.

Explain why $\int_{\text{tlow}}^{\text{thigh}} y[t]\, x'[t]\, dt$ gives you the negative area measurement.

Answer: Take another look at the plot. It starts at:

In[11]:=
```
{x[tlow],y[tlow]}
```

Out[11]=
```
{5, 0}
```

It ends at:

In[12]:=
```
{x[thigh],y[thigh]}
```

Out[12]=
```
{5, 0}
```

And the curve reverses its direction at:

In[13]:=
```
trev = Pi; {x[trev],y[trev]}
```

Out[13]=
```
{-5, 0}
```

Take a look:

In[14]:=
```
Show[redlips,
Epilog->{PointSize[0.05],
Point[{x[trev],y[trev]}]}];
```

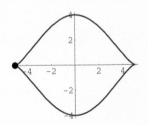

And note that

$$\int_{\text{tlow}}^{\text{thigh}} y[t]\, x'[t]\, dt = \int_{\text{tlow}}^{\text{trev}} y[t]\, x'[t]\, dt + \int_{\text{trev}}^{\text{thigh}} y[t]\, x'[t]\, dt.$$

Now look into what the two individual integrals on the right measure. The plot associated with $\int_{\text{tlow}}^{\text{trev}} y[t]\, x'[t]\, dt$ is:

In[15]:=
```
ParametricPlot[
{x[t],y[t]},{t,tlow,trev},
AspectRatio->Automatic,
PlotStyle->{{Red,Thickness[0.01]}}];
```

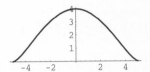

Now look at:

In[16]:=
```
{x[tlow],x[trev]}
```

Out[16]=
```
{5, -5}
```

This and the plot tell you that

$$\int_{\text{tlow}}^{\text{trev}} y[t]\, x'[t]\, dt = \int_{5}^{-5} y\, dx = -\int_{-5}^{5} y\, dx$$

is the negative of the area under the top part of the curve and over the x-axis. The plot associated with $\int_{\text{trev}}^{\text{thigh}} y[t]\, x'[t]\, dt$ is:

In[17]:=
```
ParametricPlot[
{x[t],y[t]},{t,trev,thigh},
AspectRatio->Automatic,
PlotStyle->{{Red,Thickness[0.01]}}];
```

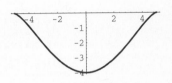

Now look at:

In[18]:=
```
{x[trev],x[thigh]}
```

Out[18]=
```
{-5, 5}
```

This and the plot tell you that

$$\int_{\text{trev}}^{\text{thigh}} y[t]\, x'[t]\, dt = \int_{-5}^{5} y\, dx$$

is the negative of the area between the bottom part of the curve and the x-axis. The negative sign comes from the fact that y is negative on the bottom part of the curve. Consequently, when you calculate

$$\int_{\text{tlow}}^{\text{thigh}} y[t]\, x'[t]\, dt = \int_{\text{tlow}}^{\text{trev}} y[t]\, x'[t]\, dt + \int_{\text{trev}}^{\text{thigh}} y[t]\, x'[t]\, dt,$$

you get the negative of the area enclosed by the entire curve.

T.3.a.iii)

> Does this work for other closed curves as well?
>
> What do you have to watch out for?

Answer: Yes, it does work for other closed curves.

If you have a closed curve given in parametric form by $\{x[t], y[t]\}$ with tlow $\leq$ $t \leq$ thigh, then $\int_{\text{tlow}}^{\text{thigh}} y[t]\, x'[t]\, dt$ measures the negative of the area enclosed by the curve provided that the curve is traced out in the counterclockwise direction as t advances from tlow to thigh.

On the other hand, $\int_{\text{tlow}}^{\text{thigh}} y[t]\, x'[t]\, dt$ measures the actual area enclosed by the curve provided that the curve is traced out in the clockwise direction as t advances from tlow to thigh. This is not a big issue because you can always fudge the minus sign.

Two other issues are important:

$\rightarrow$ None of this works if any segment of the curve is traced out more than once by $\{x[t], y[t]\}$ as t advances from tlow to thigh. Reason: You don't want to measure any part of the area more than once.

$\rightarrow$ None of this works if the curve has inner loops like this:

In[19]:=
```
Clear[t,x,y,r]; tlow = 0; thigh = 2 Pi;
r[t_] = 3.1 - 0.6 Cos[t] + 0.3 Cos[2 t] -
0.8 Cos[3 t] + 1.9 Sin[t] - 2.7 Sin[2t] + 1.8 Sin[3t];
x[t_] = r[t] Cos[t];y[t_] = r[t] Sin[t];
redlips = ParametricPlot[{x[t],y[t]},{t,tlow,thigh},
AspectRatio->Automatic,
PlotStyle->{{Red,Thickness[0.01]}}];
```

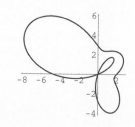

Reason: The area enclosed by the inner loop will be counted twice.

T.3.b)   This problem appears only in the electronic version.

■ T.4) Polar plots and area measurements

The usual way of specifying a point in the plane is to give its coordinates $\{x, y\}$.

In[20]:=
```
p = {1.2,1.6};
point = Graphics[{Red,PointSize[0.03],Point[p]}];
label = Graphics[Text[{"x","y"},p,{0,-3}]];
Show[point,label,Axes->True,AxesOrigin->{0,0},
PlotRange->{{0,2},{0,2}},AxesLabel->{"x","y"},
AspectRatio->Automatic];
```

You can specify the same point with the indicated polar angle t and the distance r from the point to the origin:

In[21]:=
```
dist = Graphics[{Blue,Line[{{0,0},p}]}];
rlabel = Graphics[Text["r",p/2,{0,-4}]];
angle =Graphics[Circle[{0,0},0.40,
{0,ArcTan[p[[2]]/p[[1]]]}]];
anglelabel = Graphics[Text["t",{0.45,0.12}]];
Show[point,label, dist,rlabel,angle,anglelabel,
Axes->True,AxesOrigin->{0,0},
PlotRange->{{0,2},{0,2}},AxesLabel->{"x","y"},
AspectRatio->Automatic];
```

Most folks call $\{r, t\}$ by the name the "polar coordinates of the point $\{x, y\}$." Here the polar angle t is measured in the counterclockwise sense from the x-axis and r is the distance from the origin to the point $\{x, y\}$.

If you know the polar coordinates r and t, then you can find where the point $\{x, y\}$ is: You just leave the origin $\{0, 0\}$ in the direction specified by t and you walk r units out until you get to the point $\{x, y\}$. The usual convention is: If $r > 0$, then you walk forward, but if $r < 0$, then you walk backward. If $r = 0$, then you stay put at the origin $\{0, 0\}$. In fact, if you know r and t, then you know x and y through the easy formulas:

$$x = r \cos[t]$$

and

$$y = r \sin[t].$$

T.4.a) Here's a plot of the polar curve $r[t] = 1 - \sin[t]$:

In[22]:=
```
Clear[x,y,r,t]; tlow = 0; thigh = 2 Pi;
r[t_] = 1 - Sin[t]; x[t_] = r[t] Cos[t];
y[t_] = r[t] Sin[t];
cardioid = ParametricPlot[
{x[t],y[t]},{t,tlow,thigh},
PlotStyle->{{Red,Thickness[0.01]}},
AxesLabel->{"x","y"}];
```

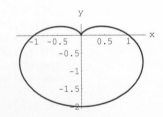

Folks like to call this curve a "cardioid" because it looks something like a heart.

> Measure the area enclosed by this curve.

Answer: You treat this as you would any closed parametric curve. Check to see whether any segment of this curve is traced out more than once by $\{x[t], y[t]\}$ as t advances from tlow to thigh:

In[23]:=
```
tmid = (tlow + thigh)/2;
ParametricPlot[
{x[t],y[t]},{t,tlow,tmid},
PlotStyle->{{Thickness[0.01],Red}},
AxesLabel->{"x","y"},AxesOrigin->{0,0}];
```

In[24]:=
```
ParametricPlot[
{x[t],y[t]},{t,tmid,thigh},
PlotStyle->{{Thickness[0.01],Red}},
AxesLabel->{"x","y"},AxesOrigin->{0,0}];
```

Hunky-dory. No segment of this curve is traced out more than once by $\{x[t], y[t]\}$ as t advances from tlow to thigh. Now calculate $\int_{\text{tlow}}^{\text{thigh}} y[t]\, x'[t]\, dt$:

In[25]:=
```
NIntegrate[y[t] x'[t],{t,tlow,thigh}]
```

Out[25]=
```
-4.71239
```

The negative sign tells you that the curve was traced out in the counterclockwise manner. The area enclosed by the curve measures out to $3\pi/2$ square units.

T.4.b.i) Here's a plot of the polar curve $r[t] = 1 - 2\cos[t]$:

In[26]:=
```
Clear[x,y,r,t]; tlow = 0; thigh = 2 Pi;
r[t_] = 1 - 2 Cos[t]; x[t_] = r[t] Cos[t];
y[t_] = r[t] Sin[t];
limacon = ParametricPlot[
{x[t],y[t]},{t,tlow,thigh},
PlotStyle->{{Blue,Thickness[0.01]}},
AxesLabel->{"x","y"}];
```

Folks like to call this curve a "limaçon" because it looks like a lima bean.

Measure the area enclosed by this curve.

Answer: Careful, you don't want to count the area inside the inner loop. The inner loop comes from negative $r[t]$'s. To get an idea of what's happening, plot $r[t]$:

In[27]:=
```
Plot[r[t],{t,tlow,thigh}];
```
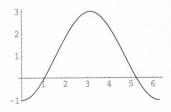

Check to see when $r[t] = 0$.

In[28]:=
```
Solve[r[t] == 0,t]
```
Out[28]=
$$\{\{t \rightarrow \frac{Pi}{3}\}\}$$

In[29]:=
```
outertlow = Pi/3; outerthigh = 2 Pi - Pi/3;
outercurve = ParametricPlot[
{x[t],y[t]},{t,outertlow,outerthigh},
PlotStyle->{{Blue,Thickness[0.01]}},
AxesLabel->{"x","y"},Epilog->Text[
"Outer Curve",{-1.5,0.5}]];
```
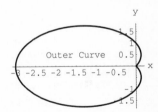

Good. Now check to see whether any segment of the outer curve is traced out more than once as t advances from outertlow to outerthigh:

In[30]:=
```
outertmid = (outertlow + outerthigh)/2;
ParametricPlot[{x[t],y[t]},
{t,outertlow,outertmid},
PlotStyle->{{Thickness[0.01],Blue}},
AxesLabel->{"x","y"},AxesOrigin->{0,0}];
```
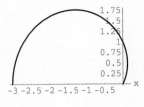

In[31]:=
```
ParametricPlot[{x[t],y[t]},
{t,outertmid,outerthigh},
PlotStyle->{{Thickness[0.01],Blue}},
AxesLabel->{"x","y"},AxesOrigin->{0,0}];
```
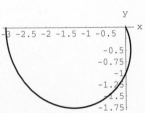

Copacetic. No segment of the outer curve is traced out more than once by $\{x[t], y[t]\}$ as t advances from outertlow to outerthigh. Now calculate

$$\int_{\text{outertlow}}^{\text{outerthigh}} y[t]\, x'[t]\, dt :$$

In[32]:=
```
outerarea = NIntegrate[y[t] x'[t],{t,outertlow,outerthigh}]
```

Out[32]=
```
   -8.88126
```

The negative sign tells you that the outer curve was traced out in the counterclockwise manner. The area enclosed by the outer curve measures out to 8.88126 square units.

T.4.b.ii) This problem appears only in the electronic version.

■ T.5) Gauss's normal law

There are many students out there who attempt to deal with Gauss's normal law in probability or statistics courses that make no use of calculus. Trying to do this is like scrubbing a floor with a toothbrush. Pity those poor suckers. There are also plenty of students who try to make calculations like those below by hand. They spend most of their time calculating. Pity them as well. You are in the catbird's seat because you have calculus and *Mathematica* working for you.

T.5.a.i) The average number of hours that a certain light bulb shines before it fails is 3210 hours. This measurement is normally distributed with a standard deviation of 207 hours.

Estimate a number b so that

$$\int_{3210-b}^{3210+b} \text{normal}[x, 3210, 207]\, dx \geq 0.95$$

and interpret the result.

Answer: Crank up the function normal[x, mean, dev]:

In[33]:=
```
Clear[x,bell]; Clear[normal,mean,dev]
bell[x_] = (1/Sqrt[2 Pi]) E^((-x^2)/2);
normal[x_,mean_,dev_] = bell[(x - mean)/dev]/dev
```

Out[33]=
```
                      1
    -------------------------------------
                      2      2
            (-mean + x) /(2 dev )
    dev E                          Sqrt[2 Pi]
```

Put

$$f[b] = \int_{3210-b}^{3210+b} \text{normal}[x, 3210, 207]\, dx$$

and plot:

In[34]:=
```
Clear[f,b]
f[b_] := NIntegrate[normal[x,3210,207],
{x,3210 - b,3210 + b},AccuracyGoal->2];
Plot[{0.95,f[b]},{b,100,500},
PlotStyle->{{Red},{Blue,Thickness[0.01]}},
AxesLabel->{"b","f[b]"}];
```

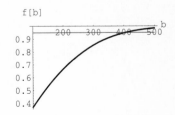

Put a microscope on it:

In[35]:=
```
Plot[{0.95,f[b]},{b,400,420},
PlotStyle->{{Red},
{Blue,Thickness[0.01]}},
AxesLabel->{"b","f[b]"}];
```

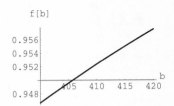

$b = 406$ will work well:

In[36]:=
```
f[406]
```

Out[36]=
```
0.950162
```

Now you know that with $b = 406$,

$$f[b] = \int_{3210-b}^{3210+b} \text{normal}[x, 3210, 207]\, dx > 0.95.$$

The interpretation:

$\rightarrow$ 95% of the light bulbs will burn out after they have shone between

$$3210 - 406 = 2804 \text{ hours}$$

and

$$3210 + 406 = 3616 \text{ hours}.$$

Statistics folks would say that the life of one of these light bulbs is between 2804 hours and 3616 hours with confidence level 95%.

T.5.a.ii) What fraction of the light bulbs described above are likely to burn out before they have shone 2500 hours?

Answer: Calculate

$$\int_0^{2500} \text{normal}[x, 3210, 207]\, dx :$$

In[37]:=
```
NIntegrate[normal[x,3210,207],{x,0,2500},AccuracyGoal->2]
```

Out[37]=
```
0.000301777
```

Only 0.03% of the bulbs are likely to burn out before they have shone 2500 hours.

Give It a Try

Experience with the starred ($\star$) problems will be especially beneficial for understanding later lessons.

■ G.1) Transforming integrals*

G.1.a) Here is *Mathematica*'s calculation of

$$\int_a^b e^{\cos[x]} \sin[x] \, dx :$$

In[1]:=
```
Clear[x,a,b]
MathematicaCalculation = Integrate[(E^Cos[x]) Sin[x],{x,a,b}]
```

Out[1]=
```
 Cos[a]      Cos[b]
E       - E
```

> Use a transformation to explain where this result comes from.

G.1.b) Here is *Mathematica*'s calculation of

$$\int_a^b e^{\cos[x^2]} \sin[x^2] \, x \, dx :$$

In[2]:=
```
Clear[x,a,b]
MathematicaCalculation = Integrate[(E^Cos[x^2]) Sin[x^2] x,{x,a,b}]
```

Out[2]=
```
      2           2
 Cos[a ]     Cos[b ]
E           E
--------- - ---------
    2           2
```

> Use a transformation to explain where this result comes from.

G.1.c) Here is *Mathematica*'s calculation of

$$\int_a^b \frac{2\,x}{1+x^2}\,dx :$$

In[3]:=
```
Clear[x,a,b]
MathematicaCalculation = Integrate[2 x /(1 + x^2),{x,a,b}]
```

Out[3]=
```
         2            2
-Log[1 + a ] + Log[1 + b ]
```

Use a transformation to explain where this result comes from.

G.1.d) Here is *Mathematica*'s calculation of

$$\int_a^b \sin[2\,x]^4 \, \cos[2\,x]\,dx :$$

In[4]:=
```
Clear[x,a,b]
MathematicaCalculation = Integrate[Sin[2 x]^4 Cos[2 x],{x,a,b}]
```

Out[4]=
```
            5            5
 -Sin[2 a]      Sin[2 b]
 ---------  +  ---------
    10            10
```

Use a transformation to explain where this result comes from.

■ G.2) Transforming integrals to explain *Mathematica* output[★]

One of the problems with old-fashioned calculus courses is that they tend to deal with only a very restricted collection of functions. Beyond calculus, important yet unfamiliar functions lurk, just waiting to participate in science. The most important of these is Erf[x]. Here's a plot of Erf[x]:

In[5]:=
```
Clear[x]
Plot[{-1,1,Erf[x]},{x,-5,5},
PlotStyle->{{Red},{Red},
{Thickness[0.01]}},
AxesLabel->{"x","Erf[x]"}];
```

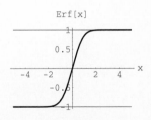

Erf[x] is defined by

$$\text{Erf}[x] = \left(\frac{2}{\sqrt{\pi}}\right) \int_0^x e^{-t^2}\, dt.$$

Now look at *Mathematica*'s calculation of

$$\int_0^b e^{-x^2/2}\, dx :$$

In[6]:=
```
Clear[x,b]
Integrate[E^((-x^2)/2),{x,0,b}]
```

Out[6]=
```
       Pi        b
Sqrt[--] Erf[-------]
       2       Sqrt[2]
```

Explaining where this result comes from is the same as explaining why

$$\int_0^b e^{-x^2/2}\, dx = \sqrt{\frac{\pi}{2}}\ \text{Erf}\left[\frac{b}{\sqrt{2}}\right]$$

$$= \sqrt{\frac{\pi}{2}}\ \left(\frac{2}{\sqrt{\pi}}\right) \int_0^{b/\sqrt{2}} e^{-t^2}\, dt$$

$$= \sqrt{2} \int_0^{b/\sqrt{2}} e^{-t^2}\, dt.$$

To do this, you want to go with a substitution that transforms $\int_0^b e^{-x^2/2}\, dx$ into $\sqrt{2}\int_0^{b/\sqrt{2}} e^{-t^2}\, dt$. The natural substitution here is

$$t = \frac{x}{\sqrt{2}}$$

(which is the same as $x = \sqrt{2}\, t$). This gives the pairings

$$e^{-x^2/2} \longleftrightarrow e^{-(t\sqrt{2})^2/2} = e^{-t^2};$$

$$dx \longleftrightarrow \sqrt{2}\, dt;$$

$$\int_0^b \longleftrightarrow \int_0^{b/\sqrt{2}}$$

because $t = 0$ when $x = 0$ and $t = b/\sqrt{2}$ when $x = b$. This gives

$$\int_0^b e^{-x^2/2}\, dx = \sqrt{2} \int_0^{b/\sqrt{2}} e^{-t^2}\, dt$$

and the explanation is complete.

G.2.a.i)   Here is *Mathematica*'s calculation of

$$\int_0^{\pi/2} e^{-(\sin[x]^2)}\ \cos[x]\, dx :$$

In[7]:=
```
Clear[x]
Integrate[E^(-Sin[x]^2) Cos[x],{x,0,Pi/2}]
```

Out[7]=
$$\frac{\text{Sqrt[Pi] Erf[1]}}{2}$$

Remembering that

$$\text{Erf}[x] = \left(\frac{2}{\sqrt{\pi}}\right) \int_0^x e^{-t^2}\, dt,$$

use a transformation to explain where this result comes from.

G.2.a.ii) Here is *Mathematica*'s calculation of

$$\int_a^{a+s} e^{-((x-a)/s)^2}\, dx :$$

In[8]:=
```
Clear[x,a,s]
Integrate[E^-((x - a)/s)^2,{x,a,a + s}]
```

Out[8]=
$$\frac{\text{Sqrt[Pi] s Erf[1]}}{2}$$

Remembering that

$$\text{Erf}[x] = \left(\frac{2}{\sqrt{\pi}}\right) \int_0^x e^{-t^2}\, dt,$$

use a transformation to explain where this result comes from.

How do you explain the fact that if you hold s constant but change the value of a, then the value of

$$\int_a^{a+s} e^{-((x-a)/s)^2}\, dx$$

does not change?

G.2.b) The gamma function is defined for $s > 0$ by

$$\text{Gamma}[s] = \int_0^\infty x^{s-1} e^{-x}\, dx.$$

Some folks like the gamma function so much that the gamma function is programmed into *Mathematica*:

In[9]:=
```
Clear[s,x]
Integrate[x^(s - 1) E^(-x),{x,0,Infinity}]
```
Out[9]=
```
Gamma[s]
```

Here's a plot of the gamma function:

In[10]:=
```
Clear[s]; b = 5;
gammaplot = Plot[Gamma[s],{s,0,b},
PlotStyle->{{Blue,Thickness[0.01]}},
AxesLabel->{"s","Gamma[s]"},
PlotRange->{0,Gamma[b]}];
```

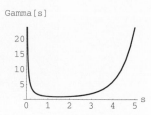

Here is *Mathematica*'s calculation of

$$\int_1^\infty \frac{\log[u]^k}{u^2} \, du :$$

In[11]:=
```
Clear[k,u]
Integrate[(Log[u]^(k))/u^2,{u,1,Infinity}]
```
Out[11]=
```
Gamma[1 + k]
```

> Use the transformation based on the substitution $u = e^x$ to explain where this result comes from.

G.2.c.i) Here is *Mathematica*'s calculation of

$$\int_0^x \frac{\sin[t]}{t} \, dt :$$

In[12]:=
```
Clear[x,t]
Integrate[Sin[t]/t,{t,0,x}]
```
Out[12]=
```
SinIntegral[x]
```

This is as it should be because SinIntegral[x] is defined by

$$\text{SinIntegral}[x] = \int_0^x \frac{\sin[t]}{t} \, dt.$$

Here's a plot of this unfamiliar critter:

In[13]:=
```
Plot[SinIntegral[x],{x,1,50},
PlotStyle->{{Red,Thickness[0.01]}},
PlotRange->All,AxesLabel->
{"x","SinIntegral[x]"}];
```

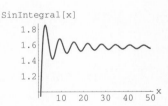

It's impossible to express this weird function in terms of more common functions. Now take a gander at *Mathematica*'s calculation of

$$\int_2^3 \frac{\sin[x^2]}{x}\, dx :$$

In[14]:=
```
Clear[x]
MathematicaCalculation = Integrate[Sin[x^2]/x,{x,2,3}]
```

Out[14]=

$$\frac{-\text{SinIntegral}[4]}{2} + \frac{\text{SinIntegral}[9]}{2}$$

Remembering that

$$\text{SinIntegral}[x] = \int_0^x \frac{\sin[t]}{t}\, dt,$$

use a transformation to explain where this result comes from.

G.2.c.ii) Here's *Mathematica*'s calculation of

$$\int_a^b \sin[e^x]\, dx :$$

In[15]:=
```
Clear[x,a,b]
MathematicaCalculation = Integrate[Sin[E^x],{x,a,b}]
```

Out[15]=

$$-\text{SinIntegral}[E^a] + \text{SinIntegral}[E^b]$$

Remembering that

$$\text{SinIntegral}[x] = \int_0^x \frac{\sin[t]}{t}\, dt,$$

use a transformation to explain where this result comes from.

G.2.d.i) Here is *Mathematica*'s calculation of

$$\int_0^x \sin\left[\frac{\pi}{2}\,t^2\right]\,dt:$$

In[16]:=
```
Clear[x,t]
Integrate[Sin[Pi/2 t^2],{t,0,x}]
```
Out[16]=
```
FresnelS[x]
```

This is as it should be because FresnelS[x] is defined by

$$\text{FresnelS}[x] = \int_0^x \sin\left[\frac{\pi}{2}\,t^2\right]\,dt.$$

It's impossible to express this weird function in terms of more common functions. Now lay your eyes on *Mathematica*'s calculation of

$$\int_0^4 \sin\left[\frac{\pi}{2}\,x^4\right]\,x\,dx:$$

In[17]:=
```
Clear[x]
MathematicaCalculation = Integrate[Sin[(Pi/2) x^4] x,{x,0,4}]
```
Out[17]=
$$\frac{\text{FresnelS}[16]}{2}$$

Remembering that

$$\text{FresnelS}[x] = \int_0^x \sin\left[\frac{\pi}{2}\,t^2\right]\,dt,$$

use a transformation to explain where this result comes from.

■ G.3) Area measurements*

G.3.a) Here's a curve coming from parametric equations:

In[18]:=
```
Clear[t,x,y]; tlow = 0; thigh = 2;
x[t_] = 7 t^2 E^(- t/2);
y[t_] = 3 Sqrt[t] (1 + Sin[2 Pi t]^2);
ParametricPlot[{x[t],y[t]},{t,tlow,thigh},
AspectRatio->Automatic,PlotStyle->
{{Red,Thickness[0.01]}},
AxesLabel->{"x","y"}];
```

Measure the area under the curve and over the x-axis.

G.3.b) Here's a closed curve coming from parametric equations:

```
In[19]:=
    Clear[t,x,y]; tlow = 0; thigh = 2 Pi;
    x[t_] = 2 Cos[t] + 4 Sin[4 t]^2;
    y[t_] = 6 Sin[t]^3;
    ParametricPlot[{x[t],y[t]},{t,tlow,thigh},
    AspectRatio->Automatic,PlotPoints->50,
    PlotStyle->{{DarkGreen,Thickness[0.01]}}];
```

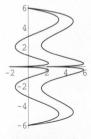

Determine whether the curve is traced out in a clockwise or counterclockwise manner as t advances from tlow to thigh.

Measure the area inside this curve.

G.3.c) Here's a curve parameterized in polar form.

```
In[20]:=
    Clear[x,y,r,t]; tlow = -Pi; thigh = Pi;
    r[t_] = 1 + 2 Sin[t] - Sin[2 t]/2;
    x[t_] = r[t] Cos[t];
    y[t_] = r[t] Sin[t];
    ParametricPlot[{x[t],y[t]},{t,tlow,thigh},
    PlotStyle->{{Blue,Thickness[0.01]}},
    AxesLabel->{"x","y"}];
```

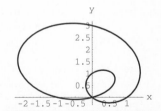

Measure the area inside the outer curve, and then measure the area inside the loop.

■ G.4) Volume measurements

This problem appears only in the electronic version.

■ G.5) Gauss's normal law all around us[*]

G.5.a.i) Calculus&Mathematica is pleased to acknowledge very helpful suggestions from Professor David Appleyard of Carleton College.

If Sally Powchinsky, age 12.3, scores a 136 on a national intelligence exam and the average 14.6-year-old scores 136, then Sally's IQ is reported to be:

In[21]:=
```
100 (14.6)/(12.3)
```
Out[21]=
```
118.699
```

Sally is smart. Years ago, IQs were trusted measurements; but nowadays the IQ measurement is taken with a good deal of skepticism. In spite of this, you will read in some education and psychology texts that intelligence quotients in the general population are normally distributed with a mean of 100 and a standard deviation of 16.

> Estimate the fraction of people with IQs between 110 and 130.

G.5.a.ii) Psychologists say that geniuses are people whose IQs measure out at 140 and higher.

> Estimate the fraction of geniuses in the whole population.

G.5.b) The idea for this problem came from the book *Statistics*, by David Freedman, Robert Pisani, Roger Purves, and Ani Adhikari, W. W. Norton, 1991.

In 1967, the math SAT (Scholastic Aptitude Test) scores were normally distributed with mean 492 and standard deviation of about 100. In 1987, the math SAT scores were normally distributed with mean 476 and standard deviation of about 100. Here's a look:

In[22]:=
```
Clear[x,bell]
bell[x_] = (1/Sqrt[2 Pi]) E^((-x^2)/2);
Clear[normal,mean,dev]
normal[x_,mean_,dev_] =
bell[(x - mean)/dev]/dev;
Plot[{normal[x,492,100],normal[x,476,100]},
{x,200,800},PlotStyle->{{Thickness[0.01]},
{Red,Thickness[0.01]}},
AxesLabel->{"Math\n SAT\n score",""}];
```

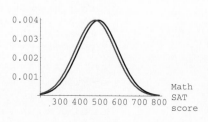

The bell for 1967 is to the right of the bell for 1987.

> Estimate the percentage of students scoring between 650 and 800 in 1967 and the percentage of students scoring between 650 and 800 in 1987.
>
> Use your results to discuss the sentiments in the following sentences:
>
> A 1967 student was almost 50% more likely to score between 650 and 800 than was a 1987 student. This is the reason that in 1967 there was more competition to get into science, engineering, and math than there was in 1987.

G.5.c.i) As you know, coin-operated coffee machines sometimes overflow the paper cup and sometimes underfill the paper cup. When a certain machine is set to dispense c ounces of coffee, the amount of coffee actually dispensed averages out to c ounces. Data collections show that the amounts it actually dispenses are normally distributed with an average of c ounces and a standard deviation of 0.35 ounces. If you set the machine to dispense exactly 7 ounces, then you expect the machine to overflow a 7-ounce cup half the time. If you don't believe this, look at

$$\int_7^\infty \text{normal}[x, 7, 0.35]\, dx:$$

In[23]:=
```
Clear[x,bell]
bell[x_] = (1/Sqrt[2 Pi]) E^((-x^2)/2);
Clear[normal,mean,dev]
normal[x_,mean_,dev_] = bell[(x - mean)/dev]/dev; mean = 7; dev = 0.35;
NIntegrate[normal[x,mean,dev],{x,mean,Infinity}]
```

Out[23]=
```
0.5
```

> If you set the machine to dispense 6.65 ounces, how often do you expect the machine to overflow a 7-ounce cup?

G.5.c.ii)

> Come up with an estimate of the largest number c such that if you set the machine to dispense c ounces, then it will overflow a 7-ounce cup no more than 0.5% of the time.

G.5.d.i) The C&M Light Bulb Company has come up with a way to make 100-watt long-life light bulbs. The average number of hours that these new light bulbs shine before they burn out is 12,678 hours. This measurement is normally distributed with standard deviation of 1113 hours.

> Estimate a number b so that
> $$\int_{12678-b}^{12678+b} \text{normal}[x, 12678, 1113]\, dx \geq 0.90$$
> and interpret the result.

■ G.6) Using transformations to analyze normally distributed measurements

American women aged 18 to 21 have an average weight of about 134 pounds. This weight measurement is normally distributed with a standard deviation of about 27 pounds. To estimate the fraction of these women whose weight is between $134 - 27$ pounds and $134 + 27$ pounds, you calculate:

In[24]:=
```
Clear[x,bell]
bell[x_] = (1/Sqrt[2 Pi]) E^((-x^2)/2);
Clear[normal,mean,dev]
normal[x_,mean_,dev_] = bell[(x - mean)/dev]/dev;
mean = 134; dev = 27;
Clear[x]
NIntegrate[normal[x,mean,dev],{x,mean - dev,mean + dev}]
```
Out[24]=
```
0.682689
```

About 68% of the weights came out between (mean − dev) and (mean + dev).

→ In 1988, female students averaged 455 on the math SAT. Their scores were approximately normally distributed with a standard deviation of about 100 points. To estimate the fraction of these women whose score was between $(450 - 100)$ points and $(450 + 100)$ points, you calculate:

In[25]:=
```
mean = 450; dev = 100;
Clear[x]
NIntegrate[normal[x,mean,dev], {x,mean - dev,mean + dev}]
```
Out[25]=
```
0.682689
```

About 68% of the women's math SAT scores came out between (mean − dev) and (mean + dev).

→ If you have any normally distributed measurement M with average = mean and standard deviation = dev, to estimate the fraction of the measurements that came out between (mean − dev) and (mean + dev), you calculate:

In[26]:=
```
Clear[x,mean,dev]
calculation = Integrate[normal[x,mean,dev],{x,mean - dev,mean + dev}];
N[calculation]
```
Out[26]=
```
0.682689
```

This tells you that when you have any normally distributed measurement M with average = mean and standard deviation = dev then about 68% of the measurements will come out between (mean − dev) and (mean + dev).

G.6.a) Part of the job of mathematics is to explain why things happen the way they do. The reason all the calculations above came out the way they did is that

$$\int_{\text{mean}-\text{dev}}^{\text{mean}+\text{dev}} \text{normal}[x, \text{mean}, \text{dev}]\, dx = \int_{-1}^{1} \text{bell}[t]\, dt :$$

In[27]:=
```
Clear[t,x,bell]; Clear[normal,mean,dev]
bell[x_] = (1/Sqrt[2 Pi]) E^((-x^2)/2);
normal[x_,mean_,dev_] = bell[(x - mean)/dev]/dev;
Integrate[normal[x,mean,dev],{x,mean-dev,mean + dev}] ==
Integrate[bell[t],{t,-1,1}]
```

Out[27]=
```
True
```

> Remembering that
>
> $$\text{normal}[x, \text{mean}, \text{dev}] = \text{bell}\left[\frac{x - \text{mean}}{\text{dev}}\right]\left(\frac{1}{\text{dev}}\right),$$
>
> use a transformation based on the substitution
>
> $$t = \frac{x - \text{mean}}{\text{dev}}$$
>
> to transform
>
> $$\int_{\text{mean}-\text{dev}}^{\text{mean}+\text{dev}} \text{normal}[x, \text{mean}, \text{dev}]\, dx \qquad \text{into} \qquad \int_{-1}^{1} \text{bell}[t]\, dt.$$

G.6.b.i) You can use the same substitution to reveal a little more. Look at this:

In[28]:=
```
Clear[s,t,x,bell]; Clear[normal,mean,dev]
bell[x_] = (1/Sqrt[2 Pi]) E^((-x^2)/2);
normal[x_,mean_,dev_] = bell[(x - mean)/dev]/dev;
Integrate[normal[x,mean,dev],
{x,mean - s dev,mean + s dev}] == Integrate[bell[t],{t,-s,s}]
```

Out[28]=
```
True
```

This tells you that

$$\int_{\text{mean}-s\,\text{dev}}^{\text{mean}-s\,\text{dev}} \text{normal}[x, \text{mean}, \text{dev}]\, dx = \int_{-s}^{s} \text{bell}[t]\, dt.$$

> Remembering that
>
> $$\text{normal}[x, \text{mean}, \text{dev}] = \text{bell}\left[\frac{x - \text{mean}}{\text{dev}}\right]\left(\frac{1}{\text{dev}}\right),$$
>
> use the substitution
>
> $$t = \frac{x - \text{mean}}{\text{dev}}$$

to transform

$$\int_{\text{mean}-s\,\text{dev}}^{\text{mean}-s\,\text{dev}} \text{normal}[x, \text{mean}, \text{dev}]\, dx \qquad \text{into} \qquad \int_{-s}^{s} \text{bell}[t]\, dt.$$

G.6.b.ii) Here's a plot of $f[s] = \int_{-s}^{s} \text{bell}[t]\, dt$ for $0 \le s \le 4$ together with the line $y = 0.5$:

```
In[29]:=
  Clear[f,s,t,x,bell]
  bell[x_] = (1/Sqrt[2 Pi]) E^((-x^2)/2);
  f[s_] = Integrate[bell[x],{x,-s,s}];
  Plot[{0.5,f[s]},{s,0,4},
  PlotStyle->{{Red},{Thickness[0.01],Blue}},
  AxesLabel->{"s","f[s]"}];
```

Get a good estimate of the special number s^* that makes

$$f[s^*] = 0.5.$$

Use the number s^* you came up with and the content of part G.6.b.i) above to explain the statement: When you have any normally distributed measurement M with average = mean and standard deviation = dev, then

$\rightarrow$ about 25% of the measurements will come out less than mean $- s^*$ dev;

$\rightarrow$ about 25% of the measurements will come out between mean $- s^*$ dev and mean;

$\rightarrow$ about 25% of the measurements will come out between mean and mean $+$ s^* dev;

$\rightarrow$ about 25% of the measurements will come out more than mean $+ s^*$ dev.

G.6.b.iii) As mentioned above, American women aged 18 to 21 have an average weight of about 134 pounds. This weight measurement is normally distributed with a standard deviation of about 27 pounds.

Use the information in part G.6.b.ii) above to give approximate weights w_1, w_2, and w_3 such that

$\rightarrow$ About 25% of these women weigh less than w_1 pounds;

$\rightarrow$ About 25% of these women weigh between w_1 and w_2 pounds;

$\rightarrow$ About 25% of these women weigh between w_2 and w_3 pounds;

$\rightarrow$ About 25% of these women weigh more than w_3 pounds.

G.6.c) As mentioned above, female students averaged 455 on the 1988 math SAT. Their scores were approximately normally distributed with a standard deviation of about 100 points.

Give approximate scores $w_1, w_2, w_3, w_4, w_5, w_6, w_7, w_8$, and w_9 such that

$\rightarrow$ About 10% of these women scored less than w_1 points;

$\rightarrow$ About 10% of these women scored between w_1 and w_2 points;

$\rightarrow$ About 10% of these women scored between w_2 and w_3 points;

$\rightarrow$ About 10% of these women scored between w_3 and w_4 points;

$\rightarrow$ About 10% of these women scored between w_4 and w_5 points;

$\rightarrow$ About 10% of these women scored between w_5 and w_6 points;

$\rightarrow$ About 10% of these women scored between w_6 and w_7 points;

$\rightarrow$ About 10% of these women scored between w_7 and w_8 points;

$\rightarrow$ About 10% of these women scored between w_8 and w_9 points;

$\rightarrow$ About 10% of these women scored more than w_9 points.

■ G.7) Transforming integrals to explain measurements of area, length and volume

This problem appears only in the electronic version.

■ G.8) A transformation bails out a materials science student

This is a true story that happened at the University of Illinois on September 16, 1992.

When material scientists measure the total energy density of photons inside an isothermal oven, they encounter the integral

$$\int_0^\infty \frac{x^3}{e^{tx} - 1}\, dx \qquad \text{where } t > 0.$$

One student at Illinois ran into this integral in his materials science homework. After searching through the CRC Tables and finding nothing that would help, he went all over the math department looking for someone who could calculate this integral by hand. No professor or student could help him; so he stopped in at the Calculus&*Mathematica* lab. In the lab, one of his friends typed:

In[30]:=
```
Clear[x,t]
result = Integrate[x^3/(E^(t x) - 1),{x,0,Infinity}]
```

Out[30]=

$$\frac{Pi^4}{15\,t^4}$$

The materials science student said: "Fine, but how do I explain this to my professor in my homework write-up?" Lots of C&M students got involved, and a big mathematical bull session ignited. Finally someone said: "Would you be happy with an explanation of this result?"

In[31]:=
```
numericalresult = N[result]
```

Out[31]=

$$\frac{6.49394}{t^4}$$

The student said: "Of course."

The reply: "Use the substitution $s = t\,x$ to transform the original integral

$$\int_0^\infty \frac{x^3}{e^{tx} - 1}\, dx \qquad \text{into} \qquad \left(\frac{1}{t^4}\right) \int_0^\infty \frac{s^3}{e^s - 1}\, ds."$$

G.8.a.i) | Do it. |

G.8.a.ii) The lab person continued: "Now you can see that

$$\int_0^\infty \frac{x^3}{e^{tx} - 1}\, dx = \left(\frac{1}{t^4}\right) \int_0^\infty \frac{s^3}{e^s - 1}\, ds$$

is given by:"

In[32]:=
```
Clear[s]
(1/t^4) NIntegrate[s^3/(E^s - 1),{s,0,Infinity}]
```

Out[32]=

$$\frac{6.49394}{t^4}$$

Compare:

In[33]:=
```
numericalresult
```

Out[33]=

$$\frac{6.49394}{t^4}$$

The materials science student thanked the bull session participants and went on his way, asking himself why he hadn't taken advantage of C&M when he took calculus.

Use the same idea to come up with and to explain a numerical formula for

$$\int_0^\infty \frac{x^2}{e^{tx} - 1}\, dx \qquad \text{for } t > 0.$$

■ G.9) Gauss's normal law and grading on the curve

This problem appears only in the electronic version.

■ G.10) Polar plots: Rotating and measuring

G.10.a) Here's an ellipse parameterized by $\{x[t], y[t]\}$ with $0 \le t \le 2\pi$:

```
In[34]:=
  Clear[x,y,r,t]; x[t_] = 5 Cos[t];
  y[t_] = 3 Sin[t]; tlow = 0; thigh = 2 Pi;
  original = ParametricPlot[
  {x[t],y[t]},{t,tlow,thigh},
  PlotStyle->{{Blue,Thickness[0.02]}},
  AspectRatio->Automatic,AxesLabel->{"x","y"}];
```

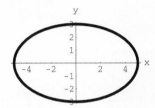

You can rotate the curve s radians in the counterclockwise direction by using

$$\{\text{xrot}[t], \text{yrot}[t]\} = \{\cos[s]\, x[t] - \sin[s]\, y[t], \sin[s]\, x[t] + \cos[s]\, y[t]\}.$$

Here's what this ellipse looks like when it is rotated by $s = \pi/6$ radians in the counterclockwise direction:

```
In[35]:=
  Clear[xrot,yrot,s]
  xrot[t_] = Cos[s] x[t] - Sin[s] y[t];
  yrot[t_] = Sin[s] x[t] + Cos[s] y[t];
  s = Pi/6; rotated = ParametricPlot[
  {xrot[t],yrot[t]},{t,tlow,thigh},
  PlotStyle->{{Red,Thickness[0.01]}},
  AxesLabel->{"x","y"},AspectRatio->Automatic];
```

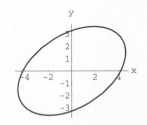

Here they are together:

```
In[36]:=
  Show[original,rotated,
  AspectRatio->Automatic];
```

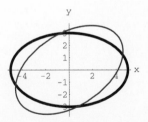

Here's what you get when you measure the area inside the original ellipse:

In[37]:=
```
Integrate[y[t] x'[t],{t,tlow,thigh}]
```

Out[37]=
```
-15 Pi
```

And the measurement of the area inside the rotated ellipse:

In[38]:=
```
Integrate[yrot[t] xrot'[t],{t,tlow,thigh}]
```

Out[38]=
```
-15 Pi
```

The area inside the original ellipse measures out to the same number as the area in the rotated ellipse.

> Why isn't this a surprise?

G.10.a.ii) Here's another parametric curve given in polar form:

In[39]:=
```
Clear[x,y,r,t]; tlow = 0; thigh = 2 Pi;
r[t_] = 4  + 3 Sin[3 t];
x[t_] = r[t] Cos[t]; y[t_] = r[t] Sin[t];
original = ParametricPlot[
{x[t],y[t]},{t,tlow,thigh},
PlotStyle->{{Blue,Thickness[0.02]}},
AspectRatio->Automatic,
AxesLabel->{"x","y"}];
```

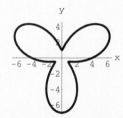

> Rotate this closed curve by $\pi/4$ radians in the counterclockwise sense and show the original curve and the rotated curve together.
>
> Measure the area enclosed by each curve.

G.10.a.iii) Here's another parametric curve given in polar form:

In[40]:=
```
Clear[x,y,r,t]; tlow = 0; thigh = 2 Pi;
r[t_] = 3 + Sin[t] + Sin[2 t] + Sin[4 t];
x[t_] = r[t] Cos[t];y[t_] = r[t] Sin[t];
original = ParametricPlot[
{x[t],y[t]},{t,tlow,thigh},
PlotStyle->{{Blue,Thickness[0.02]}},
AspectRatio->Automatic,AxesLabel->{"x","y"}];
```

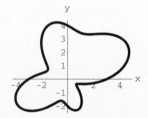

> Rotate this closed curve by $\pi/2$ radians in the counterclockwise sense and show
> the original curve and the rotated curve together. Measure the area enclosed
> by each curve.

G.10.b.i) Actually, when you take any function $r[t]$ with $r[t]$ never negative, then the curve
parameterized by $\{x[t], y[t]\} = \{r[t]\cos[t], r[t]\sin[t]\}$, with $0 \le t \le 2\pi$, always
traces out in the counterclockwise manner as t advances from 0 to 2π. In addition,
because $r[t]$ is never negative, you are always guaranteed that the resulting curve
has no loops and no segment is ever repeated. This tells you that if you take:

In[41]:=
```
Clear[x,y,r,t]
{x[t_],y[t_]} = {r[t] Cos[t],r[t] Sin[t]}
```
Out[41]=
```
{Cos[t] r[t], r[t] Sin[t]}
```

To measure the area inside the curve, you can integrate: $y[t]\,x'[t]$ from 0 to 2π:

In[42]:=
```
integrand1 = Expand[y[t] D[x[t],t],Trig->True]
```
Out[42]=
$$-\frac{r[t]^2}{2} + \frac{\cos[2\ t]\ r[t]^2}{2} + \frac{r[t]\ \sin[2\ t]\ r'[t]}{2}$$

Ouch; integrand1 is a mess. You can also choose a countertclockwise rotation angle
$s = \pi/2$ and put:

In[43]:=
```
Clear[xrot,yrot]; s = Pi/2;
{xrot[t_],yrot[t]} = {Cos[s] x[t] - Sin[s] y[t], Sin[s] x[t] + Cos[s] y[t]}
```
Out[43]=
```
{-(r[t] Sin[t]), Cos[t] r[t]}
```

To measure the area inside the curve, you can integrate: $\text{yrot}[t]\,\text{xrot}'[t]$ from 0 to
2π:

In[44]:=
```
integrand2 = Expand[yrot[t] D[xrot[t],t],Trig->True]
```
Out[44]=
$$-\frac{r[t]^2}{2} - \frac{\cos[2\ t]\ r[t]^2}{2} - \frac{r[t]\ \sin[2\ t]\ r'[t]}{2}$$

Neither integrand1 nor integrand2 is a thing of beauty:

In[45]:=
```
{integrand1,integrand2}
```

Out[45]=

$$\left\{\frac{-r[t]^2}{2} + \frac{\text{Cos}[2\ t]\ r[t]^2}{2} + \frac{r[t]\ \text{Sin}[2\ t]\ r'[t]}{2}, \right.$$

$$\left. \frac{-r[t]^2}{2} - \frac{\text{Cos}[2\ t]\ r[t]^2}{2} - \frac{r[t]\ \text{Sin}[2\ t]\ r'[t]}{2}\right\}$$

Now see what you get when you average integrand1 and integrand2:

In[46]:=
```
integrand3 = (integrand1 + integrand2)/2
```

Out[46]=

$$\frac{-r[t]^2}{2}$$

That's dandy.

> Explain why you can use the nifty formula
>
> $$\left(\frac{1}{2}\right) \int_0^{2\pi} r[t]^2\, dt$$
>
> to measure the area inside the curve.

G.10.b.ii) Here's yet another polar curve:

In[47]:=
```
Clear[x,y,r,t]; tlow = 0; thigh = 2 Pi;
r[t_] = 5 + 3 Cos[3 t] - 1.5 Sin[5 t];
x[t_] = r[t] Cos[t];y[t_] = r[t] Sin[t];
original = ParametricPlot[
{x[t],y[t]},{t,tlow,thigh},
PlotStyle->{{Blue,Thickness[0.02]}},
AspectRatio->Automatic,AxesLabel->{"x","y"}];
```

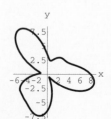

As t advances from 0 to 2π, this curve plots out in the counterclockwise manner with no segment repeated.

> Explain why you can calculate the area inside this curve by calculating
>
> $$-\int_0^{2\pi} y[t]x'[t]dt :$$
>
> Or by calculating
>
> $$\int_0^{2\pi} \frac{r[t]^2}{2}dt.$$

■ G.11) Counterclockwise or clockwise?★

G.11.a.i) Here's the circle $x^2 + y^2 = 4$ plotted via the parametric equations $x[t] = 2\cos[t]$ and $y[t] = 2\sin[t]$ with $0 \le t \le 2\pi$:

In[48]:=
```
Clear[x,y,t]; tlow = 0; thigh = 2 Pi;
x[t_] = 2 Cos[t]; y[t_] = 2 Sin[t];
original = ParametricPlot[
{x[t],y[t]},{t,tlow,thigh},
PlotStyle->{{Blue,Thickness[0.02]}},
AxesLabel->{"x","y"},AspectRatio->Automatic];
```

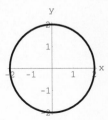

Here is the measurement of the area inside the circle:

In[49]:=
```
originalarea = Integrate[y[t] x'[t],{t,tlow,thigh}]
```

Out[49]=
```
-4 Pi
```

Here is the same circle $x^2 + y^2 = 4$ plotted via the parametric equations $xx[t] = 2\sin[t]$ and $yy[t] = 2\cos[t]$ with $0 \le t \le 2\pi$:

In[50]:=
```
Clear[xx,yy,t]; tlow = 0; thigh = 2 Pi;
xx[t_] = 2 Sin[t]; yy[t_] = 2 Cos[t];
new = ParametricPlot[
{xx[t],yy[t]},{t,tlow,thigh},
PlotStyle->{{Blue,Thickness[0.02]}},
AxesLabel->{"x","y"},AspectRatio->Automatic];
```

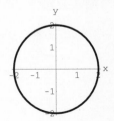

Here is the measurement of the area inside the circle:

In[51]:=
```
newarea = Integrate[yy[t] xx'[t],{t,tlow,thigh}]
```

Out[51]=
```
4 Pi
```

Look at:

In[52]:=
```
{originalarea,newarea}
```

Out[52]=
```
{-4 Pi, 4 Pi}
```

How do these area measurements reveal which parameterization was clockwise and which was counterclockwise?

G.11.a.ii) Here is a curve plotted from parametric equations:

In[53]:=
```
Clear[x,y,t]; tlow = 0; thigh = 2 Pi;
x[t_] = 4 t Sin[t/2]^2;
y[t_] = 4 Sin[t] + 2 Sin[2 t];
ParametricPlot[
 {x[t],y[t]},{t,tlow,thigh},
 PlotStyle->{{Blue,Thickness[0.01]}},
 AxesLabel->{"x","y"},AspectRatio->Automatic];
```

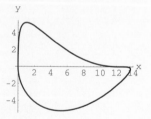

Now look at a measurement of the area inside the curve:

In[54]:=
```
NIntegrate[y[t] x'[t],{t,tlow,thigh}]
```

Out[54]=
```
78.9568
```

> Use the sign of the result immediately above to answer the question: As t advances from tlow to thigh, the parameterization $\{x[t], y[t]\}$ traces out the curve in which manner: clockwise or counterclockwise?

■ G.12) Work and velocity

A force $f[x]$ imparts an acceleration $a[x]$ to an object of constant mass m moving (with no friction) along the x-axis from $x = a$ to $x = b$. Physicists have a notion of the work done by this force while it moves the object from $x = a$ to $x = b$. They measure work with an integral:

$$\text{Work} = \int_a^b m\, a[x]\, dx.$$

G.12.a)
> Why is
> $$a[x] = v[x]\, v'[x]$$
> where $v[x]$ is velocity at position x for $a \leq x \leq b$?

G.12.b)
> Evaluate the integral
> $$\text{Work} = \int_a^b m\, a[x]\, dx$$
> by replacing $a[x]$ by
> $$v[x]\, v'[x]$$

to get

$$\text{Work} = \int_a^b m\,v[x]\,v'[x]\,dx$$

and then using the pairings

$$m\,v[x] \longleftrightarrow mv$$
$$v'[x]\,dx \longleftrightarrow dv$$
$$\int_a^b \longleftrightarrow \int_{v[a]}^{v[b]}$$

to get a formula for work in terms of the velocities $v[a]$ at $x = a$ and $v[b]$ at $x = b$.

The formula that you get will show that the work measurement done does not depend on the explicit nature of the force. The force may vary in magnitude in any imaginable way provided that the velocities at the two endpoints, $v[a]$ and $v[b]$, do not change. In other words, the work measurement depends only on the mass and the beginning velocity $v[a]$ and the terminal velocity $v[b]$.

LESSON 2.05

2D Integrals and the Gauss-Green Formula

Basics

■ B.1) 2D integrals for volume measurements

Here's a plot of the function $f[x] = e^{-0.2x} \cos[4\,x]$:

```
In[1]:=
  Clear[f,x]; {a,b} = {0,4};
  f[x_] = E^(-0.2 x) Cos[4 x];
  fplot = Plot[f[x],{x,a,b},
  PlotStyle->{{Blue,Thickness[0.01]}},
  AxesLabel->{"x","y"}];
```

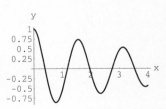

To get the points on the plot, you go to a point $\{x_0, 0\}$ on the x-axis underneath or above the plot and then you run a line to the point $\{x_0, f[x_0]\}$ on the curve like this:

```
In[2]:=
  xo = 1.5; Show[fplot,
  Graphics[{PointSize[0.02],Point[{xo,0}]}],
  Graphics[{PointSize[0.02],Point[{xo,f[xo]}]}],
  Graphics[{Red,Line[{{xo,0},{xo,f[xo]}}]}]];
```

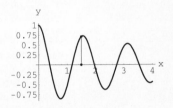

Play with this by resetting x_0 and rerunning. You can get a pretty good idea of what the curve looks like by just looking at the tips of the little line segments:

195

In[3]:=
```
xjump = (b - a)/50;
Show[fplot,Table[
{Graphics[{PointSize[0.02],Point[{x,0}]}],
Graphics[{PointSize[0.02],Point[{x,f[x]}]}],
Graphics[{Red,Line[{{x,0},
{x,f[x]}}]}]}],{x,a,b,xjump}]];
```

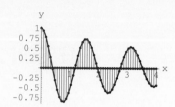

You can plot a surface $z = f[x, y]$ like this:

In[4]:=
```
Clear[f,x,y]; f[x_,y_] = 3.1 x^2 + 2.3 y^2;
{{a,b},{c,d}} = {{-2,3},{-1,4}};
surfaceplot = Plot3D[f[x,y],{x,a,b},{y,c,d},
DisplayFunction->Identity];
spacer = 0.2; threedims = Graphics3D[{
{Blue,Line[{{a,0,0},{b,0,0}}]},
Text["x",{b + spacer,0,0}],{Blue,Line[{{0,c,0},{0,d,0}}]},
Text["y",{0,d + spacer,0}],{Blue,Line[{{0,0,0},{0,0,50}}]},
Text["z",{0,0,50 + spacer}]}];CMView = {2.7,1.6,1.2};
fplot = Show[surfaceplot,threedims,ViewPoint->CMView,
PlotRange->All,DisplayFunction->$DisplayFunction];
```

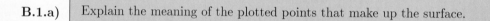

B.1.a)   Explain the meaning of the plotted points that make up the surface.

Answer: You do it just as you did it above. To get the points on the plot, you go to a point $\{x_0, y_0, 0\}$ on the xy-plane underneath or above the surface. Then you run a line to the 3D point $\{x_0, y_0, f[x_0, y_0]\}$ like this:

In[5]:=
```
{xo,yo} = {2.5,3};
Show[fplot,Graphics3D[{PointSize[0.02],
Point[{xo,yo,0}]}],
Graphics3D[{PointSize[0.02],
Point[{xo,yo,f[xo,yo]}]}],
Graphics3D[{Red,Line[{{xo,yo,0},
{xo,yo,f[xo,yo]}}]}]]];
```

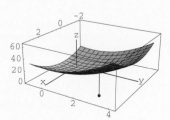

Play with this by resetting x_0 and y_0 and rerunning. You can get a pretty good idea of what the surface looks like by just looking at the tips of the little line segments:

In[6]:=
```
xjump = (b - a)/8; yjump = (d - c)/8;
Show[fplot,Table[{Graphics3D[
{PointSize[0.02],Point[{x,y,0}]}],
Graphics3D[{PointSize[0.02],
Point[{x,y,f[x,y]}]}],
Graphics3D[{Red,Line[{{x,y,0},{x,y,f[x,y]}}]}]},
{x,a,b,xjump},{y,c,d,yjump}]];
```

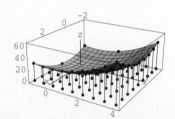

Think of the line segments as poles propping up the surface.

B.1.b.i) Take another look at the plot from part B.1.a) immediately above:

> How do you use the double integral $\iint_R f[x,y]\,dx\,dy$ to measure the volume of the solid consisting of everything between the surface plotted above and the xy-plane?

Answer: Look again at the poles plotted above:

In[7]:=
```
xjump = (b - a)/10; yjump = (d - c)/10;
Show[fplot,Table[{Graphics3D[
{PointSize[0.02],Point[{x,y,0}]}],
Graphics3D[{Red,Line[{{x,y,0},
{x,y,f[x,y]}}]}]}],{x,a,b,xjump},
{y,c,d,yjump}]];
```

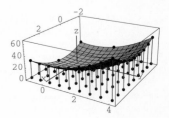

The solid region you're talking about consists of all the poles you get when you plot all the possible poles (not just those plotted above). When you plotted this surface on the top, you ran x from a to b and you ran y from c to d. The volume of the solid is given by the double integral

$$\iint_R f[x,y]\,dx\,dy$$

where R is the rectangle in the xy-plane consisting of all points

$\{x,y,0\}$ with $a \le x \le b$ and $c \le y \le d$.

You can calculate the volume measurement

$$\iint_R f[x,y]\,dx\,dy$$

by writing it as

$$\int_c^d \int_a^b f[x,y]\,dx\,dy,$$

which you get from *Mathematica* with:

In[8]:=
```
Integrate[f[x,y],{x,a,b},{y,c,d}]
```

Out[8]=
```
430.
```

The volume of the solid measures out to 430 cubic units.

B.1.b.ii)   What is the physical process that confirms that the double integral

$$\iint_R f[x,y]\, dx\, dy = \int_c^d \int_a^b f[x,y]\, dx\, dy$$

actually measures the volume of the solid described in part B.1.b.i)?

Answer: Look at some slices of the solid consisting of everything between the surface plotted above and the xy-plane:

In[9]:=
```
Clear[slice,x,y,t]
slice[y_] := ParametricPlot3D[{x,y,0} + t {0,0,f[x,y]},
{x,a,b},{t,0,1},PlotPoints->{Automatic,2},DisplayFunction->Identity];
jump = (d - c)/3; Table[Show[fplot,slice[y],
DisplayFunction->$DisplayFunction],{y,c,d,jump}];
```

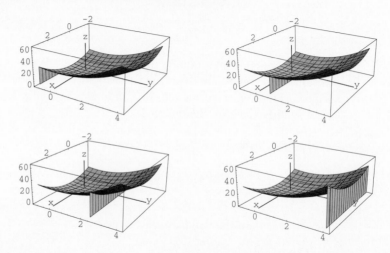

The physical process of calculating $\int_c^d \int_a^b f[x,y]\, dx\, dy$ involves two steps.

First step: For each y with $c \leq y \leq d$, you calculate $\int_a^b f[x,y]\, dx$:

In[10]:=
```
Clear[first,x,y]; first[y_] = Integrate[f[x,y],{x,a,b}]
```

Out[10]=
$$36.1667 + 11.5\, y^2$$

As y advances from c to d, first$[y]$ measures the area of the slices that sweep out the solid. For instance,

In[11]:=
```
first[3]
```

Out[11]=
```
139.667
```

measures the area of this slice of the solid:

In[12]:=
```
Show[fplot,slice[3],
DisplayFunction->$DisplayFunction];
```

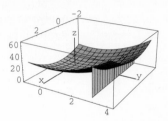

Second step: Calculate

$$\int_c^d \text{first}[y]\, dy = \int_c^d \int_a^b f[x,y]\, dx\, dy.$$

This second integral sums up the area measurements of the individual slices and measures the volume.

In[13]:=
```
second = Integrate[first[y],{y,c,d}]
```

Out[13]=
```
430.
```

Compare to *Mathematica*'s calculation of $\int_c^d \int_a^b f[x,y]\, dx\, dy$:

In[14]:=
```
Integrate[f[x,y],{y,c,d},{x,a,b}]
```

Out[14]=
```
430.
```

Bingo.

■ B.2) What $\iint_R f[x,y]\, dx\, dy$ means when $f[x,y]$ isn't always positive

B.2.a) Here's a calculation of $\int_0^4 \int_0^2 f[x,y]\, dx\, dy$ for $f[x,y] = (x-3)\, e^{0.43y}$:

In[15]:=
```
Clear[f,x,y]
f[x_,y_] = (x - 3) E^(0.43 y);
Integrate[f[x,y],{x,0,2},{y,0,4}]
```

Out[15]=
```
-42.6468
```

Negative. When the dweeby Calculus Cal looked at this, he said, "This has got to be wrong because double integrals measure volume and volume can't be negative."

What do you say to Cal in response?

Answer: You realize that Cal's question does have scientific merit and you answer it by telling Cal to look at the integral

$$\int_0^4 \int_0^2 f[x,y]\, dx\, dy$$

and the corresponding plot of $f[x,y]$:

```
In[16]:=
  {a,b} = {0,2}; {c,d} = {0,4};
  fplot = Plot3D[f[x,y],{x,a,b},{y,c,d},
  DisplayFunction->Identity];
  spacer = 0.2; threedims = Graphics3D[{
  {Blue,Line[{{a,0,0},{b,0,0}}]},
  Text["x",{b + spacer,0,0}],{Blue,Line[{{0,c,0},{0,d,0}}]},
  Text["y",{0,d + spacer,0}],{Blue,Line[{{0,0,-10},{0,0,5}}]},
  Text["z",{0,0,5 + spacer}]}]; CMView = {2.7,1.6,1.2};
  Show[fplot,threedims,ViewPoint->CMView,
  DisplayFunction->$DisplayFunction];
```

You say that the plot reveals that $f[x,y]$ is negative for the x's and y's involved in the calculation of

$$\int_0^4 \int_0^2 f[x,y]\, dx\, dy.$$

This is why

$$\int_0^4 \int_0^2 f[x,y]\, dx\, dy$$

is negative. You go on to tell Cal that if he wants a measurement of the volume of the solid whose *bottom* skin is the plotted part of the surface and whose *top* skin is the rectangle in the xy-plane consisting of the points $\{x,y,0\}$ with $0 \le x \le 2$ and $0 \le y \le 4$, then he can get the measurement by calculating

$$-\int_0^4 \int_0^2 f[x,y]\, dx\, dy :$$

```
In[17]:=
  -Integrate[f[x,y],{x,0,2},{y,0,4}]

Out[17]=
  42.6468
```

Positive. Now Cal is happy and so are you.

B.2.b.i)   Here's a calculation of

$$\int_0^4 \int_0^5 f[x,y]\, dx\, dy \qquad \text{for } f[x,y] = (x-3)\, e^{0.43y} :$$

```
In[18]:=
  Clear[f,x,y]
  f[x_,y_] = (x - 3) E^(0.43 y); Integrate[f[x,y],{x,0,5},{y,0,4}]
```

Out[18]=
-26.6542

Try to interpret this calculation.

Answer: Look at the integral $\int_0^4 \int_0^5 f[x,y] \, dx \, dy$ and the corresponding plot of $f[x,y]$:

In[19]:=
```
{a,b} = {0,5}; {c,d} = {0,4};
fplot = Plot3D[f[x,y],{x,a,b},{y,c,d},
DisplayFunction->Identity]; spacer = 0.2;
threedims = Graphics3D[{
{Blue,Line[{{a,0,0},{b,0,0}}]},
Text["x",{b + spacer,0,0}],
{Blue,Line[{{0,c,0},{0,d,0}}]},
Text["y",{0,d + spacer,0}],{Blue,Line[{{0,0,-10},{0,0,5}}]},
Text["z",{0,0,5 + spacer}]}]; CMView = {2.7,1.6,1.2};
Show[fplot,threedims,ViewPoint->CMView,
DisplayFunction->$DisplayFunction];
```

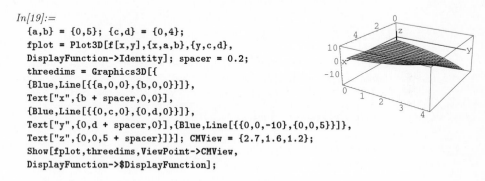

Look at the plot to see that $f[x,y]$ is sometimes positive and sometimes negative as x varies between 0 and 5 and y varies between 0 and 4. The upshot:

$$\int_0^4 \int_0^5 f[x,y] \, dx \, dy$$

does not measure any specific volume.

B.2.b.ii) Now you know that when you go with $f[x,y] = (x-3) \, e^{0.43y}$, then you can calculate $\int_0^4 \int_0^5 f[x,y] \, dx \, dy$.

In[20]:=
```
Clear[f,x,y]
f[x_,y_] = (x - 3) E^(0.43 y); Integrate[f[x,y],{x,0,5},{y,0,4}]
```

Out[20]=
-26.6542

But the integral doesn't measure any specific volume because $f[x,y]$ takes on both positive and negative values as x varies between 0 and 5 and y varies between 0 and 4.

Just how can you interpret this integral?

Answer: Look at:

In[21]:=
```
f[x,y]
```

Out[21]=

$$\underset{E}{0.43\ y} \quad (-3 + x)$$

You can see that $f[x, y] \geq 0$ for $x \geq 3$ and $f[x, y] \leq 0$ for $x \leq 3$. To get the interpretation of $\int_0^4 \int_0^5 f[x, y]\, dx\, dy$, write

$$\int_0^4 \int_0^5 f[x, y]\, dx\, dy = \int_0^4 \int_0^3 f[x, y]\, dx\, dy + \int_0^4 \int_3^5 f[x, y]\, dx\, dy.$$

→ In the first integral, $f[x, y] \geq 0$; so,

$$\int_0^4 \int_0^3 f[x, y]\, dx\, dy$$

measures the volume of the solid whose *bottom* skin is the rectangle in the xy-plane consisting of all points $\{x, y, 0\}$ with $0 \leq x \leq 3$ and $0 \leq y \leq 4$ and whose *top* skin is the surface $z = f[x, y]$ plotted with $0 \leq x \leq 3$ and $0 \leq y \leq 4$.

→ In the second integral, $f[x, y] \leq 0$; so,

$$\int_0^4 \int_3^5 f[x, y]\, dx\, dy$$

measures the *negative* of the volume of the solid whose *top* skin is the rectangle in the xy-plane consisting of all points $\{x, y, 0\}$ with $3 \leq x \leq 5$ and $0 \leq y \leq 4$ and whose *bottom* skin is the surface $z = f[x, y]$ plotted with $3 \leq x \leq 5$ and $0 \leq y \leq 4$.

Consequently,

$$\int_0^4 \int_0^5 f[x, y]\, dx\, dy = \int_0^4 \int_0^3 f[x, y]\, dx\, dy + \int_0^4 \int_3^5 f[x, y]\, dx\, dy$$

measures the difference of the volumes of the two solids calculated above.

Just a little bit esoteric.

■ B.3) Trying to calculate $\iint_R f[x, y]\, dx\, dy$ when R isn't a rectangle

B.3.a) Here is a little region R on the xy-plane.

In[22]:=
```
Clear[high,low,x]
high[x_] = 2 Sin[Pi x/3];
low[x_] = (1/2) x (x - 3);
regionplot = Plot[{high[x],low[x]},{x,0,3},
PlotStyle->{{Blue,Thickness[0.01]}},
AspectRatio->Automatic,AxesLabel->{"x","y"},
Epilog->{Text["y = high[x]",{0.7,1.5}],
Text["y = low[x]",{2.0,-0.8}],
Text["R",{1.5,1}]}];
```

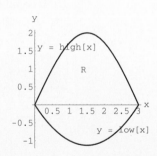

The region R consists of all points between and on the plotted curves.

> Calculate
>
> $$\iint_R f[x,y]\,dx\,dy \qquad \text{for } f[x,y] = e^{-0.4x-0.3y}.$$

Answer: Enter the function $f[x,y]$:

In[23]:=
```
Clear[f,x,y]
f[x_,y_] = E^(-0.4 x - 0.3 y);
```

To set up $\iint_R f[x,y]\,dx\,dy$ for calculation, you need a convenient way of slicing R. Here's one:

In[24]:=
```
Clear[slice,sliceplot,x]
slice[x_] := Graphics[{Red,
Line[{{x,low[x]},{x,high[x]}}]}];
sliceplot[x_] := Show[regionplot,
slice[x],DisplayFunction->Identity];
Show[sliceplot[1],
DisplayFunction->$DisplayFunction];
```

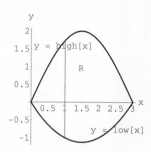

You slice from low to high, and this means that you integrate with respect to y first. This tells you how to set up

$$\iint_R f[x,y]\,dx\,dy = \int_0^3 \left(\int_{\text{low}[x]}^{\text{high}[x]} f[x,y]\,dy \right) dx$$

for calculation:

In[25]:=
```
Clear[first]
first[x_] = Integrate[f[x,y],{y,low[x],high[x]}]
```

Out[25]=
```
           -0.4 x - 0.15 (-3 + x) x              -0.4 x - 0.6 Sin[(Pi x)/3]
   3.33333 E                         - 3.33333 E
```

You can't use NIntegrate on the first integral because the formula for the first integral includes an x. You can use NIntegrate on the second integral:

In[26]:=
```
second = NIntegrate[first[x],{x,0,3}]
```

Out[26]=
```
   3.20682
```

Here's how to get *Mathematica* to calculate the double integral

$$\iint_R f[x,y]\,dx\,dy = \int_0^3 \int_{\text{low}[x]}^{\text{high}[x]} f[x,y]\,dy\,dx$$

in one sweet instruction with NIntegrate:

In[27]:=
```
NIntegrate[f[x,y],{x,0,3},{y,low[x],high[x]}]
```

Out[27]=
```
3.20682
```

Done.

B.3.b) Here is another little region on the xy-plane.

In[28]:=
```
Clear[left,right,y]; left[y_] = y;
right[y_] = 2 - y^2;
regionplot = ParametricPlot[{
{left[y],y},{right[y],y}},
{y,-2,1},PlotStyle->{{Blue,Thickness[0.01]}},
AspectRatio->Automatic,AxesLabel->{"x","y"},
Epilog->{Text["x = left[y]",{-1,-0.6}],
Text["x = right[y]",{1.5,-1}],
Text["R",{1,-0.5}]}];
```

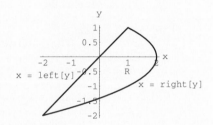

The region R consists of all points between and on the plotted curves.

> Calculate
>
> $$\iint_R f[x,y]\,dx\,dy \qquad \text{for } f[x,y] = xy.$$

Answer: Enter the function $f[x,y]$:

In[29]:=
```
Clear[f,x,y]; f[x_,y_] = x y;
```

To set up $\iint_R f[x,y]\,dx\,dy$ for calculation, you need a convenient way of slicing R. Here's one:

In[30]:=
```
Clear[slice,sliceplot,y]
slice[y_] := Graphics[{Red,
Line[{{left[y],y},{right[y],y}}]}];
sliceplot[y_] := Show[regionplot,
slice[y],DisplayFunction->Identity];
Show[sliceplot[-0.5],
DisplayFunction->$DisplayFunction];
```

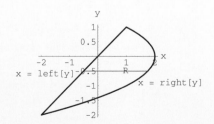

You slice from left to right, and this means that you integrate with respect to x first. This tells you how to set up

$$\iint_R f[x,y]\,dx\,dy = \left(\int_{-2}^{1} \int_{\text{left}[y]}^{\text{right}[y]} f[x,y]\,dx \right) dy$$

for calculation:

In[31]:=
```
Clear[first]
first[y_] = Integrate[f[x,y],{x,left[y],right[y]}]
```

Out[31]=

$$\frac{-y^3}{2} + \frac{y\,(2 - y^2)^2}{2}$$

You can't use NIntegrate on the first integral because the formula for the first integral includes an x. You can use NIntegrate on the second integral.

In[32]:=
```
second = Integrate[first[x],{x,-2,1}]
```

Out[32]=

$$\frac{9}{8}$$

Here's how to get *Mathematica* to calculate the double integral

$$\iint_R f[x,y]\,dx\,dy = \int_{-2}^{1} \int_{\text{left}[y]}^{\text{right}[y]} f[x,y]\,dx\,dy$$

in one sweet instruction:

In[33]:=
```
Integrate[f[x,y],{y,-2,1},{x,left[y],right[y]}]
```

Out[33]=

$$\frac{9}{8}$$

Done.

■ B.4) The Gauss-Green formula helps you calculate $\iint_R f[x,y]\,dx\,dy$

One part of the art of mathematics is knowing how to avoid miserable calculations. The Gauss-Green formula is a tool for avoiding miserable calculations. The Gauss-Green formula says this:

When you have a parameterization $\{x[t], y[t]\}$ of the boundary of a region R in the xy-plane with the extra feature that as t advances from a to b, the points $\{x[t], y[t]\}$

advance around the boundary of R exactly once in the counterclockwise way, then you are guaranteed that

$$\iint_R \left(D[n[x,y],x] - D[m[x,y],y]\right) dx\, dy = \int_a^b m[x[t],y[t]]\, x'[t] + n[x[t],y[t]]\, y'[t]\, dt.$$

It's a funny-looking formula, but don't let it intimidate you. Its weirdness gives it the utility of a crescent wrench because you can adjust it to many situations. Get to feel comfortable with this formula because you can use it to side-step a lot of miserable calculations.

B.4.a)

Use the Gauss-Green formula to help calculate the double integral

$$\iint_R \left(x^2 + y^2\right) e^{-y}\, dx\, dy$$

where R is the region inside the ellipse

$$\left(\frac{x}{2}\right)^2 + y^2 = 1.$$

Answer: You just say

$$f[x,y] = \left(x^2 + y^2\right) e^{-y}$$

with

$$m[x,y] = 0$$

and

$$n[x,y] = \int_0^x f[s,y]\, ds :$$

In[34]:=
```
Clear[x,y,n,m,t]; m[x_,y_] = 0;
f[x_,y_] = (x^2 + y^2)E^(-y);
n[x_,y_] = Integrate[f[s,y],{s,0,x}]
```

Out[34]=
$$\frac{x^3}{3\,E^y} + \frac{x\,y^2}{E^y}$$

This gives you $f[x,y] = D[n[x,y],x] - D[m[x,y],y]$:

In[35]:=
```
{D[n[x,y],x]- D[m[x,y],y], f[x,y]}
```

Out[35]=
$$\left\{\frac{x^2}{E^y} + \frac{y^2}{E^y}, \frac{x^2 + y^2}{E^y}\right\}$$

Now you know that

$$\iint_R \left(x^2 + y^2\right) e^{-y} \, dx \, dy = \iint_R f[x,y] \, dx \, dy$$

$$= \iint_R \left(D[n[x,y],x] - D[m[x,y],y]\right) \, dx \, dy$$

$$= \int_a^b m[x[t],y[t]] \, x'[t] + n[x[t],y[t]] \, y'[t] \, dt$$

where $\{x[t], y[t]\}$ is a parameterization of the ellipse $(x/2)^2 + y^2 = 1$ chosen so that the points $\{x[t], y[t]\}$ advance around the ellipse exactly once in the counterclockwise fashion as t advances from a to b.

One parameterization of $(x/2)^2 + y^2 = 1$ that does this nicely is

$$\{x[t], y[t]\} = \{2\cos[t], \sin[t]\} \qquad \text{with } a = 0 \text{ and } b = 2\pi.$$

The upshot:

$$\iint_R \left(x^2 + y^2\right) e^{-y} \, dx \, dy$$

is given by:

In[36]:=
```
a = 0; b = 2 Pi; Clear[x,y,t]
{x[t_],y[t_]} = {2 Cos[t],Sin[t]};
NIntegrate[m[x[t],y[t]] x'[t] + n[x[t],y[t]] y'[t],{t,a,b}]
```

Out[36]=
```
8.80785
```

That's all there is to it. You're out of here without worrying about slicing or anything like that.

B.4.b) To use the Gauss-Green formula to calculate $\iint_R f[x,y] \, dx \, dy$, you need a parameterization $\{x[t], y[t]\}$ of the boundary R with the extra feature that as t advances from a to b, the points $\{x[t], y[t]\}$ advance around the boundary of R exactly once in the counterclockwise way. Then you are guaranteed that when you set

$$n[x,y] = \int_0^x f[s,y] \, ds,$$

you are guaranteed that

$$\iint_R f[x,y] \, dx \, dy = \int_a^b n[x[t],y[t]] \, y'[t] \, dt.$$

The beauty of this is that it allows you to calculate a 2D integral by means of a single integral that you can usually calculate with NIntegrate.

Discuss why this works.

Answer: It's all a matter of the fundamental formula of calculus. The fundamental formula of calculus tells you that when you start with a function $f[x]$ and you put

$$n[x] = \int_0^x f[s]\, ds,$$

you are guaranteed that $D[n[x], x] = f[x]$. Try it out.

In[37]:=
```
Clear[f,n,s,x]; f[x_] = x^2;
n[x_] = Integrate[f[s],{s,0,x}]
```

Out[37]=
```
 3
x
─
3
```

Compare $D[n[x], x]$ and $f[x]$:

In[38]:=
```
{D[n[x],x],f[x]}
```

Out[38]=
```
  2   2
{x , x }
```

The same. Play with this by changing $f[x]$ and rerunning. The fundamental formula also says that when you start with a function $f[x, y]$ and you put

$$n[x, y] = \int_0^x f[s, y]\, ds$$

you are guaranteed that $D[n[x, y], y] = f[x, y]$. Try it out.

In[39]:=
```
Clear[f,n,s,x,y]; f[x_,y_] = x^2 Sin[y];
n[x_,y_] = Integrate[f[s,y],{s,0,x}]
```

Out[39]=
```
 3
x  Sin[y]
─────────
    3
```

Compare $D[n[x, y], x]$ and $f[x, y]$:

In[40]:=
```
{D[n[x,y],x],f[x,y]}
```

Out[40]=
```
  2          2
{x  Sin[y], x  Sin[y]}
```

The same. Play with this by changing $f[x, y]$ and rerunning. When you feel comfortable with this, then you are ready to use the Gauss-Green formula comfortably.

Here is the Gauss-Green formula again: When you have a parameterization $\{x[t], y[t]\}$ of the boundary of a region R in the xy-plane with the extra feature that as t advances from a to b, the points $\{x[t], y[t]\}$ advance around the boundary of R exactly once in the counterclockwise way, then you are guaranteed that

$$\iint_R D[n[x,y],x] - D[m[x,y],y]\,dx\,dy = \int_a^b m[x[t],y[t]]\,x'[t] + n[x[t],y[t]]\,y'[t]\,dt.$$

To use this formula to calculate $\iint_R f[x,y]\,dx\,dy$ for a given function $f[x,y]$, you just set

$$m[x,y] = 0 \qquad \text{for all } x\text{'s and } y\text{'s}$$

and

$$n[x,y] = \int_0^x f[s,y]\,ds.$$

This guarantees that $D[n[x,y],x] = f[x,y]$ and $D[m[x,y],y] = 0$. So,

$$\iint_R f[x,y]\,dx\,dy = \iint_R D[n[x,y],x]\,dx\,dy$$

$$= \iint_R (D[n[x,y],x] - 0)\,dx\,dy$$

$$= \iint_R (D[n[x,y],x] - D[m[x,y],y])\,dx\,dy$$

$$= \int_a^b m[x[t],y[t]]\,x'[t] + n[x[t],y[t]]\,y'[t]\,dt$$

$$= \int_a^b (0 + n[x[t],y[t]]\,y'[t])\,dt$$

$$= \int_a^b n[x[t],y[t]]\,y'[t]\,dt.$$

Explanation complete.

B.4.c) The region R in the xy-plane consists of everything inside and on the curve you see below:

In[41]:=
```
Clear[x,y,t];
{x[t_],y[t_]} = {2,3} +
{Cos[t](2 - 1.5 Cos[2 t]^2),Sin[t]};
{a,b} = {0,2 Pi}; ParametricPlot[
{x[t],y[t]},{t,a,b},PlotStyle->
{{Thickness[0.01],Blue}},AspectRatio->Automatic,
AxesLabel->{"x","y"},Epilog->Text["R",{2,3}]];
```

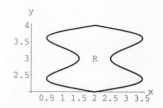

Use the Gauss-Green formula to calculate

$$\iint_R x^2\,y\,dx\,dy.$$

Answer: First check to see whether the parameterization of the boundary of R is counterclockwise:

In[42]:=
```
early = ParametricPlot[
 {x[t],y[t]},{t,a,a + 1},
 PlotStyle->{{Thickness[0.01],Blue}},
 AspectRatio->Automatic,
 AxesLabel->{"x","y"}];
```

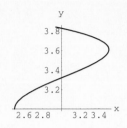

In[43]:=
```
later = ParametricPlot[
 {x[t],y[t]},{t,a + 1,a + 3},
 PlotStyle->{{Thickness[0.02],Red}},
 AspectRatio->Automatic,
 AxesLabel->{"x","y"}];
```

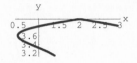

In[44]:=
```
Show[early,later];
```

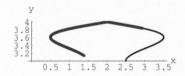

Yep; counterclockwise. Now enter $f[x, y] = x^2\, y$:

In[45]:=
```
Clear[f]; f[x_,y_] = x^2 y;
```

Set $n[x, y] = \int_0^x f[s, y]\, ds$:

In[46]:=
```
Clear[n,s]; n[x_,y_] = Integrate[f[s,y],{s,0,x}]
```

Out[46]=

$$\frac{x^3\, y}{3}$$

Calculate

$$\iint_R f[x, y]\, dx\, dy = \iint_R D[n[x, y], x]\, dx\, dy = \int_a^b n[x[t], y[t]]\, y'[t]\, dt :$$

In[47]:=
```
NIntegrate[n[x[t],y[t]] y'[t],{t,a,b}]
```

Out[47]=
```
52.706
```

No sweat.

B.4.d)   What else is the Gauss-Green formula good for?

Answer: If you're going on to vector calculus, you'll see that the Gauss-Green formula is the base of a lot of the measurement procedures you'll learn. Even if you're not going on to vector calculus, you'll see it used for a couple of snazzy things in this lesson. Stay tuned.

■ B.5)   An indication of some of the ideas behind the Gauss-Green formula

This problem appears only in the electronic version.

Tutorials

■ T.1)   Using a 2D integral to measure area

T.1.a)   Here's a little region R in the xy-plane:

```
In[1]:=
  Clear[high,low,x]
  high[x_] = 2 E^(-0.2 x) Sin[Pi x/4];
  low[x_] = 0.05 x^3 (x - 4);
  regionplot = Plot[{high[x],low[x]},
  {x,0,4},PlotStyle->{{Blue,Thickness[0.01]}},
  AspectRatio->Automatic,AxesLabel->{"x","y"},
  Epilog->{Text["y = high[x]",{3,1.1}],
  Text["y = low[x]",{3,-1.2}],
  Text["R",{2,0.75}]}];
```

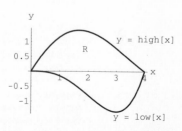

Use a single integral to measure the area of the region R between the two curves, and then use a 2D integral to make the same measurement.

Answer: Without 2D integrals, you just calculate

$$\int_0^4 \left(\text{high}[x] - \text{low}[x]\right) dx :$$

```
In[2]:=
  NIntegrate[high[x] - low[x],{x,0,4}]
```

```
Out[2]=
  6.02594
```

With 2D integrals you set $f[x, y] = 1$ for all x's and y's and calculate

$$\iint_R f[x, y]\, dx\, dy = \int_0^4 \int_{\text{low}[x]}^{\text{high}[x]} f[x, y]\, dy\, dx :$$

In[3]:=
```
Clear[y]
NIntegrate[1,{x,0,4},{y,low[x],high[x]}]
```

Out[3]=
```
6.02594
```

The reason this works is that when you go with $f[x, y] = 1$ for all x's and y's, then

$$\iint_R f[x, y]\, dx\, dy$$

measures the volume of the flat solid one unit high whose vertical sides have the same shape as R:

In[4]:=
```
Clear[r]; top = ParametricPlot3D[
{x,low[x] + r (high[x] - low[x]),1},
{x,0,4},{r,0,1},PlotPoints->{Automatic,2},
DisplayFunction->Identity];
bottom = ParametricPlot3D[
{x,low[x] + r (high[x] - low[x]),0},
{x,0,4},{r,0,1},PlotPoints->{Automatic,2},
DisplayFunction->Identity];
side1 = ParametricPlot3D[{x,low[x],r},
{x,0,4},{r,0,1},PlotPoints->{Automatic,2},
DisplayFunction->Identity];
side2 = ParametricPlot3D[{x,high[x],r},
{x,0,4},{r,0,1},PlotPoints->{Automatic,2},
DisplayFunction->Identity]; CMView = {2.7,1.6,1.2};
Show[top,bottom,side1,side2,ViewPoint->CMView,
DisplayFunction->$DisplayFunction];
```

Because this solid is one unit high, the volume measurement of the solid also gives you the area measurement of its base.

The upshot: When you go with $f[x, y] = 1$ for all x's and y's, then $\iint_R f[x, y]\, dx\, dy$ measures the area of R.

■ T.2) Volume measurements with 2D integrals

T.2.a.i) Here's a plot of the surface

$$z = f[x, y] = 12 - x + 2\, y$$

above the rectangle R in the xy-plane consisting of all points $\{x, y, 0\}$ with $-1 \le x \le 3$ and $-1 \le y \le 2$:

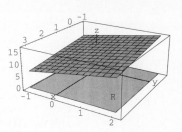

In[5]:=
```
Clear[f,x,y];
f[x_,y_] = 12 - x + 2 y;
{{a,b},{c,d}} = {{-1,3},{-1,2}};
surfaceplot = Plot3D[f[x,y],{x,a,b},{y,c,d},
DisplayFunction->Identity];
Rplot = Graphics3D[Polygon[{
{a,c,0},{a,d,0},{b,d,0},{b,c,0}}]];
label = Graphics3D[Text["R",{2,1.5,0}]];
spacer = 0.2;
threedims = Graphics3D[{{Blue,Line[{{a,0,0},{b,0,0}}]},
Text["x",{b + spacer,0,0}],{Blue,Line[{{0,c,0},{0,d,0}}]},
Text["y",{0,d + spacer,0}],{Blue,Line[{{0,0,0},{0,0,16}}]},
Text["z",{0,0,16 + 3 spacer}]}];
CMView = {2.7,1.6,1.2};
Show[surfaceplot,Rplot,label,threedims,
ViewPoint->CMView,PlotRange->All,
DisplayFunction->$DisplayFunction];
```

Measure the volume of the solid consisting of all points between or on the rectangle and the surface plotted above.

Answer: That's duck soup. The top surface is just a plot of $f[x,y]$ for points $\{x,y,0\}$ in the rectangle R in the xy-plane. The rectangle R consists of those points $\{x,y,0\}$ with $a \le x \le b$ and $c \le y \le d$. The volume measurement is just

$$\int_c^d \int_a^b f[x,y]\, dx\, dy :$$

In[6]:=
```
Integrate[f[x,y],{y,c,d},{x,a,b}]
```

Out[6]=
```
144
```

That's all there is to it.

T.2.a.ii)   Here's a plot of the surface

$$z = f[x,y] = 5 - 0.3\left(x^2 + y^2\right)$$

above region R in the xy-plane consisting of all points $\{x,y,0\}$ with $\{x,y\}$ inside the ellipse

$$\left(\frac{x}{3}\right)^2 + \left(\frac{y-1}{2}\right)^2 = 1 :$$

```
In[7]:=
  Clear[f,x,y,r,t]
  {x[r_,t_],y[r_,t_]} = {0,1} + r {3 Cos[t],2 Sin[t]};
  f[x_,y_] = 5 - 0.3 (x^2 + y^2);
  surfaceplot = ParametricPlot3D[{x[r,t],y[r,t],
  f[x[r,t],y[r,t]]},{t,0,2 Pi},
  {r,0,1},DisplayFunction->Identity];
  Rplot = ParametricPlot3D[{x[r,t],
  y[r,t],0},{t,0,2 Pi},{r,0,1},
  PlotPoints->{Automatic,2},
  DisplayFunction->Identity];
  label = Graphics3D[Text["R",{-1,2,0}]];
  spacer = 0.2; threedims = Graphics3D[{
  {Blue,Line[{{-3,0,0},{3.5,0,0}}]},
  Text["x",{3.5 + spacer,0,0}],{Blue,Line[{{0,-1,0},{0,3.5,0}}]},
  Text["y",{0,3.5 + spacer,0}],{Blue,Line[{{0,0,0},{0,0,5.5}}]},
  Text["z",{0,0,5.5 + spacer}]}]; CMView = {2.7,1.6,1.2};
  Show[surfaceplot,Rplot,label,threedims,
  ViewPoint->CMView,PlotRange->All,
  DisplayFunction->$DisplayFunction];
```

> Measure the volume of the solid consisting of all points between or on the ellipse and the surface plotted above.

Answer: The top surface is just a plot of $f[x, y]$ for points $\{x, y, 0\}$ inside the ellipse R in the xy-plane. The volume measurement is just $\iint_R f[x, y]\, dx\, dy$, which you can easily calculate with the Gauss-Green formula once you remember the ellipse

$$\left(\frac{x}{3}\right)^2 + \left(\frac{y-1}{2}\right)^2 = 1,$$

(which is the boundary of R) is parameterized in the counterclockwise way by:

```
In[8]:=
  Clear[x,y,t]
  {x[t_],y[t_]} = {0,1} + {3 Cos[t],2 Sin[t]}
Out[8]=
  {3 Cos[t], 1 + 2 Sin[t]}
```

with t running from 0 to 2π. Now you are ready to turn loose the winning team of *Mathematica* and the Gauss-Green formula: The Gauss-Green formula says

$$\iint_R D[n[x, y], x] - D[m[x, y], y]\, dx\, dy = \int_a^b m[x[t], y[t]]\, x'[t] + n[x[t], y[t]]\, y'[t]\, dt$$

where $a = 0$ and $b = 2\pi$.

To calculate

$$\iint_R f[x, y]\, dx\, dy,$$

you just go with $m[x, y] = 0$ and $n[x, y] = \int_0^x f[s, y]\, ds$:

In[9]:=
```
Clear[n,m,s]; m[x_,y_] = 0;
n[x_,y_] = Integrate[f[s,y],{s,0,x}]
```
Out[9]=
```
       3           2
 -0.1 x  + x (5 - 0.3 y )
```

This gives you $f[x, y] = D[n[x, y], x] - D[m[x, y], x]$:

In[10]:=
```
Expand[{D[n[x,y],x] - D[m[x,y],x], f[x,y]}]
```
Out[10]=
```
            2         2          2         2
 {5 - 0.3 x  - 0.3 y , 5 - 0.3 x  - 0.3 y }
```

Now you know that

$$\iint_R f[x, y] \, dx \, dy = \int_a^b m[x[t], y[t]] \, x'[t] + n[x[t], y[t]] \, y'[t] \, dt$$

where $\{x[t], y[t]\}$ is the parameterization of the boundary of R used above.

In[11]:=
```
a = 0; b = 2 Pi;
NIntegrate[m[x[t],y[t]] x'[t] + n[x[t],y[t]] y'[t],{t,a,b}]
```
Out[11]=
```
70.2146
```

The volume measures out to about 70.2 cubic units; that's all there is to it.

■ T.3) Calculation strategies for minimum effort on your part

T.3.a) Here's a region R plotted in the xy-plane.

In[12]:=
```
Clear[x]
Rplot = Plot[{x (x - 2)E^x,
x (2 - x)},{x,0,2},PlotStyle->
{{Red,Thickness[0.01]}},
AxesLabel->{"x","y"},
Epilog->Text["R",{1.25,-1.5}]];
```

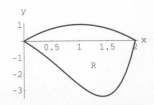

> How does this plot and its coding signal to you the minimum effort way of calculating $\iint_R f[x, y] \, dx \, dy$ for a given function $f[x, y]$?

Answer: Things go best when you slice perpendicularly to the x-axis. In fact, when you look at the code above you can see that when you slice perpendicularly to the x-axis at a fixed x, then

$$x (x - 2) \, e^x \le y \le x (2 - x) :$$

In[13]:=
```
Clear[yhigh,ylow,x,y]
yhigh[x_] = x (2 - x);
ylow[x_] = x (x - 2) E^x;
Clear[xslicer]; xslicer[x_] = Graphics[
Line[{{x,ylow[x]},{x,yhigh[x]}}]];
Show[Rplot,xslicer[0.8]];
```

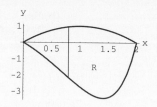

This tells you to integrate with respect to y first to get

$$\iint_R f[x,y] \, dx \, dy = \iint_R f[x,y] \, dy \, dx = \int_0^2 \int_{ylow[x]}^{yhigh[x]} f[x,y] \, dy \, dx$$

because x runs from 0 to 2. Here is the calculation in the case $f[x,y] = 5x + y$:

In[14]:=
```
Clear[f,x,y]
f[x_,y_] = 5 x + y;
first = Integrate[f[x,y],{y,ylow[x],yhigh[x]}]
```

Out[14]=

$$5 (2 - x) x^2 + \frac{(2 - x)^2 x^2}{2} - 5 E^x (-2 + x) x^2 - \frac{E^{2x} (-2 + x)^2 x^2}{2}$$

In[15]:=
```
second = NIntegrate[first,{x,0,2}]
```

Out[15]=
```
25.8908
```

Or you can try to get it with one instruction:

In[16]:=
```
N[Integrate[f[x,y],{x,0,2},{y,ylow[x],yhigh[x]}]]
```

Out[16]=
```
25.8908
```

Slicing perpendicularly to the y-axis would have resulted in a calculational nightmare; so you wouldn't want to try to integrate with respect to x first. Using the Gauss-Green formula would have been somewhat inconvenient because you would have to do a bit of thinking to come up with the required counterclockwise parameterization of the boundary.

T.3.b) Here's another region R plotted in the xy-plane.

In[17]:=
```
Clear[y]
Rplot = ParametricPlot[{{y (y - 3),y},
{y (3 - y) Sin[y]^2,y}},{y,0,3},
PlotStyle->{{Red,Thickness[0.01]}},
AxesLabel->{"x","y"},
Epilog->Text["R",{0.3,1}]];
```

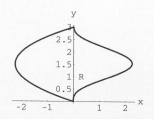

How does this plot and its code signal to you the minimum effort way of calculating $\iint_R f[x, y]\, dx\, dy$ for a given function $f[x, y]$?

Answer: You get a calculationally clean answer when you slice perpendicularly to the y-axis. In fact, when you look at the code above you can see that when you slice perpendicularly to the y-axis at a fixed y, then

$$y\,(y - 3) \le x \le y\,(3 - y)\sin[y]^2.$$

Take a look:

In[18]:=

```
Clear[xlow,xhigh,y]
xhigh[y_] = y (3 - y) Sin[y]^2;
xlow[y_] = y (y - 3); Clear[yslicer]
yslicer[y_] = Graphics[Line[
{{xlow[y],y},{xhigh[y],y}}]];
Show[Rplot,yslicer[2.3]];
```

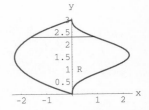

This tells you to integrate with respect to y first to get

$$\iint_R f[x, y]\, dx\, dy = \int_0^3 \int_{\text{xlow}[y]}^{\text{xhigh}[y]} f[x, y]\, dx\, dy$$

because y runs from 0 to 3. Here is the calculation in the case $f[x, y] = \sin[x + y]$:

In[19]:=

```
Clear[f,x,y]
f[x_,y_] = Sin[x + y];
first = Integrate[f[x,y],{x,xlow[y],xhigh[y]}]
```

Out[19]=

$$\text{Cos}[y + (-3 + y)\ y] - \text{Cos}[y + (3 - y)\ y\ \text{Sin}[y]^2]$$

In[20]:=

```
second = NIntegrate[first,{y,0,3}]
```

Out[20]=

```
3.02982
```

Or you can try a single instruction:

In[21]:=

```
N[Integrate[f[x,y],{y,0,3},{x,xlow[y],xhigh[y]}]]
```

Out[21]=

```
3.02982
```

Slicing perpendicularly to the x-axis would have resulted in a calculational nightmare; so you wouldn't want to try to integrate with respect to y first. Using the Gauss-Green formula would have been inconvenient because you would have to do a bit of thinking to come up with the required counterclockwise parameterization of the boundary.

T.3.c) Here's another region R plotted in the xy-plane.

In[22]:=
```
Clear[x,y,t]; x[t_] = Cos[t]
(2  + Cos[t] - Cos[2 t]/6); y[t_] = Sin[t]
(1.5 - Sin[t] + Sin[3 t]/8);
Rplot = ParametricPlot[{x[t],y[t]},{t,0, 2 Pi},
PlotRange->All,PlotStyle->{{Red,Thickness[0.01]}},
AxesLabel->{"x","y"},Epilog->Text["R",{1,-1}]];
```

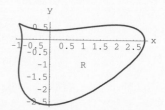

How does this plot and its code signal to you the minimum effort way of calculating

$$\iint_R f[x,y]\,dx\,dy$$

for a given function $f[x,y]$?

Answer: The boundary of R is given parametrically, and that's what Gauss-Green is set up to use. Before you use it, you have to check that the boundary curve is swept out exactly once in the counterclockwise direction.

In[23]:=
```
ParametricPlot[{x[t],y[t]},{t,0,Pi/2},
PlotRange->All,PlotStyle->
{{Red,Thickness[0.01]}},
AxesLabel->{"x","y"}];
```

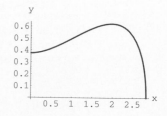

In[24]:=
```
ParametricPlot[{x[t],y[t]},{t,Pi/2,Pi},
PlotRange->All,PlotStyle->
{{Red,Thickness[0.01]}},
AxesLabel->{"x","y"}];
```

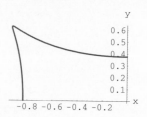

Yep, exactly once and counterclockwise. Here's how to use the Gauss-Green formula to calculate $\iint_R f[x,y]\,dx\,dy$ in the case that

$$f[x,y] = x - 2\,y:$$

The Gauss-Green formula says

$$\iint_R D[n[x,y],x] - D[m[x,y],y]\, dx\, dy = \int_a^b m[x[t],y[t]]\, x'[t] + n[x[t],y[t]]\, y'[t]\, dt$$

where $a = 0$ and $b = 2\pi$. To calculate $\iint_R (x - 2y)\, dx\, dy$, you just say

$$f[x,y] = x - 2y,$$
$$m[x,y] = 0,$$

and

$$n[x,y] = \int_0^x f[s,y]\, ds :$$

In[25]:=
```
Clear[x,y,n,m,s]; f[x_,y_] = x - 2 y;
m[x_,y_] = 0; n[x_,y_] = Integrate[f[s,y],{s,0,x}]
```

Out[25]=
$$\frac{x^2}{2} - 2\ x\ y$$

This gives you $f[x,y] = D[n[x,y],x] - D[m[x,y],x]$:

In[26]:=
```
{D[n[x,y],x] - D[m[x,y],x], f[x,y]}
```

Out[26]=
```
{x - 2 y, x - 2 y}
```

Now you know that

$$\iint_R f[x,y]\, dx\, dy = \int_a^b m[x[t],y[t]]\, x'[t] + n[x[t],y[t]]\, y'[t]\, dt$$

where $\{x[t], y[t]\}$ is the parameterization of the boundary of R used above.

In[27]:=
```
a = 0; b = 2 Pi; Clear[x,y,t]
x[t_] = Cos[t]; (2 + Cos[t] - Cos[2 t]/6);
y[t_] = Sin[t]; (1.5 - Sin[t] + Sin[3 t]/8);
NIntegrate[m[x[t],y[t]] x'[t] + n[x[t],y[t]] y'[t],{t,a,b}]
```

Out[27]=
```
21.5248
```

That's it. If you don't go with the Gauss-Green formula, you've got to slice perpendicularly to the x-axis or the y-axis. Either way, you're going to have trouble because of the bulges and hollows on the boundary of R.

T.3.d)   Given that the region R consists of everything inside the ellipse

$$\left(\frac{x-1}{5}\right)^2 + \left(\frac{y+2}{3}\right)^2 = 1$$

in the xy-plane and

$$f[x, y] = y \sin[x],$$

how does the set-up signal to you the minimum effort way of calculating $\iint_R f[x, y] \, dx \, dy$?

Answer: This is a natural for the Gauss-Green formula because you can write down a quick parameterization of the ellipse

$$\left(\frac{x-1}{5}\right)^2 + \left(\frac{y+2}{3}\right)^2 = 1.$$

Just go with the parameterization below and see the ellipse unfold right before your eager eyes:

In[28]:=
```
Clear[x,y,t]
{x[t_],y[t_]} = {1 + 5 Cos[t],-2 + 3 Sin[t]};
Rplot = ParametricPlot[{x[t],y[t]},{t,0, 2 Pi},
PlotStyle->{{Red,Thickness[0.01]}},
AxesLabel->{"x","y"}];
```

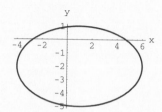

Here's how to use the Gauss-Green formula to calculate $\iint_R f[x, y] \, dx \, dy$ in the case that $f[x, y] = y \sin[x]$: The Gauss-Green formula says

$$\iint_R D[n[x, y], x] - D[m[x, y], y] \, dx \, dy = \int_a^b m[x[t], y[t]] \, x'[t] + n[x[t], y[t]] \, y'[t] \, dt$$

where $a = 0$ and $b = 2\pi$. To calculate $\iint_R y \sin[x] \, dx \, dy$, you just say

$$f[x, y] = \sin[x\, y],$$
$$m[x, y] = 0,$$

and

$$n[x, y] = \int_0^x f[s, y] \, ds :$$

In[29]:=
```
Clear[x,y,n,m,t]; f[x_,y_] = y Sin[x];
m[x_,y_] = 0; n[x_,y_] = Integrate[f[t,y],{t,0,x}]
```
Out[29]=
```
y - y Cos[x]
```

This gives you $f[x, y] = D[n[x, y], x] - D[m[x, y], y]$:

In[30]:=
```
{D[n[x,y],x]- D[m[x,y],y], f[x,y]}
```
Out[30]=
```
{y Sin[x], y Sin[x]}
```

Now you know that

$$\iint_R f[x,y]\,dx\,dy = \int_a^b m[x[t],y[t]]\,x'[t] + n[x[t],y[t]]\,y'[t]\,dt$$

where $\{x[t],y[t]\}$ is the parameterization of the boundary of R used above.

In[31]:=
```
a = 0; b = 2 Pi; Clear[x,y,t]
{x[t_],y[t_]} = {1 + 5 Cos[t],-2 + 3 Sin[t]};
NIntegrate[m[x[t],y[t]] x'[t] + n[x[t],y[t]] y'[t],{t,a,b},AccuracyGoal->3]
```

Out[31]=
```
10.3917
```

Quick and sweet.

T.3.e) Here's another region R plotted in the xy-plane.

In[32]:=
```
Clear[high,low,x]
high[x_] = 3 + x + Cos[2 x];
low[x_] = x Sin[x];
Rplot = Plot[{high[x],low[x]},{x,1,4},
PlotStyle->Red,AxesLabel->{"x","y"},
Epilog->{{Red,Line[
{{1,low[1]},{1,high[1]}}]},
{Red,Line[{{4,low[4]},{4,high[4]}}]},
Text["R",{3,3}]}];
```

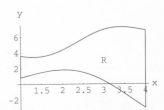

> How does this plot and its code signal to you the minimum effort way of calculating $\iint_R f[x,y]\,dx\,dy$ for a given function $f[x,y]$?

Answer: You get a calculationally clean answer when you slice perpendicularly to the x-axis. In fact, when you look at the code above you can see that when you slice perpendicularly to the y-axis at a fixed x, then $\text{low}[x] \leq y \leq \text{high}[x]$. Take a look:

In[33]:=
```
Clear[xslicer]
xslicer[x_] = Graphics[
Line[{{x,low[x]},{x,high[x]}}]];
Show[Rplot,xslicer[3.6]];
```

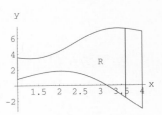

This tells you to integrate with respect to y first to get

$$\iint_R f[x,y]\,dx\,dy = \iint_R f[x,y]\,dy\,dy = \int_1^4 \int_{\text{low}[x]}^{\text{high}[x]} f[x,y]\,dy\,dx$$

because x runs from 1 to 4. Here is the calculation in the case $f[x,y] = e^{2x-y}$:

In[34]:=
```
Clear[f]; f[x_,y_] = E^(2 x - y); first = Integrate[f[x,y],{y,low[x],high[x]}]
```

Out[34]=

$$-E^{-3 + x - \text{Cos}[2\,x]} + E^{2\,x - x\,\text{Sin}[x]}$$

In[35]:=
```
second = NIntegrate[first,{x,1,4}]
```

Out[35]=
```
11140.6
```

Or you can try one instruction:

In[36]:=
```
Clear[f,x,y]; f[x_,y_] = E^(2 x - y); N[Integrate[f[x,y],{x,1,4},{y,low[x],high[x]}]]
```

Out[36]=
```
11140.6
```

Slicing perpendicularly to the *y*-axis would have resulted in a calculational nightmare; so you wouldn't want to try to integrate with respect to *x* first. Using the Gauss-Green formula would have been inconvenient because you would have to do quite a bit of work to come up with the required counterclockwise parameterization of the boundary.

■ T.4) Gauss-Green when you have a clockwise parameterization

T.4.a)

> What happens when you try to use the Gauss-Green formula but you find that your parameterization moves in the clockwise way rather than the required counterclockwise way?

Answer: Go ahead and use the clockwise parameterization to make the calculation, but adjust your answer by multiplying it by −1. The reason this works is that

$$\int_a^b g[t]\, dt = -\int_b^a g[t]\, dt.$$

■ T.5) Gauss's normal law in 2D and using it to plan bombing runs

This problem appears only in the electronic version.

Give It a Try

Experience with the starred (⋆) problems will help to increase your understanding of later material.

■ G.1) Volume measurements*

G.1.a) Here is the surface
$$z = f[x, y] = 2 \sin[x + y] + 2.5$$
plotted over the rectangle R in the xy-plane consisting of the points
$$\{x, y, 0\} \qquad \text{with} \ -2 \le x \le 2 \text{ and } -3 \le y \le 3:$$

In[1]:=
```
Clear[x,y,z,r,t,f]
f[x_,y_] = 2 Sin[x + y] + 2.5;
surfaceplot = Plot3D[f[x,y],{x,-2,2},{y,-3,3},
DisplayFunction->Identity]; Rplot = Graphics3D[
Polygon[{{-2,-3,0},{2,-3,0},{2,3,0},{-2,3,0}}]];
spacer = 0.2; threedims = Graphics3D[{
{Blue,Line[{{-2.2,0,0},{2.2,0,0}}]},
Text["x",{2.2 + spacer,0,0}],
{Blue,Line[{{0,-3.2,0},{0,3.2,0}}]},
Text["y",{0,3.2 + spacer,0}],{Blue,Line[{{0,0,0},{0,0,4.5}}]},
Text["z",{0,0,4.5 + spacer + 0.1}]}]; CMView = {2.7,1.6,1.2};
Show[threedims,Rplot,surfaceplot,PlotRange->All,
ViewPoint->CMView,Boxed->False,DisplayFunction->$DisplayFunction];
```

Use a double integral to measure the volume of the solid consisting of everything between the surface and the rectangle.

G.1.b) Here is the surface
$$z = f[x, y] = x\, y + 4$$
plotted above the region R inside the ellipse
$$\left(\frac{x}{2}\right)^2 + \left(\frac{y}{3}\right)^2 = 1$$
in the xy-plane.

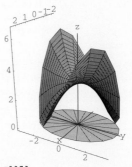

In[2]:=
```
Clear[x,y,z,r,t,f]; f[x_,y_] = x y + 3;
surfaceplot = ParametricPlot3D[{2 r Cos[t],
3 r Sin[t],f[2 r Cos[t],3 r Sin[t]]},
{r,0,1},{t,0, 2 Pi},DisplayFunction->Identity];
Rplot = ParametricPlot3D[{2 r Cos[t],
3 r Sin[t],0},{r,0,1},{t,0, 2 Pi},
PlotPoints->{2,Automatic},DisplayFunction->Identity];
threedims = Graphics3D[
{{Blue,Line[{{-2.2,0,0},{2.2,0,0}}]},
Text["x",{2.2 + spacer,0,0}],
{Blue,Line[{{0,-3.2,0},{0,3.2,0}}]},
Text["y",{0,3.2 + spacer,0}],{Blue,Line[{{0,0,0},{0,0,6.2}}]},
Text["z",{0,0,6.2 + spacer + 0.2}]}]; CMView = {2.7,1.6,1.2};
Show[threedims,surfaceplot,Rplot,PlotRange->All,
ViewPoint->CMView,Boxed->False,DisplayFunction->$DisplayFunction];
```

Think of the solid that consists of everything below the surface and above the ellipse R, and use the Gauss-Green formula to help you measure the volume of this solid.

G.1.c.i) Use the idea of volume calculation to explain the statement:

If a function $f[x, y]$ takes only positive values on a region R in the xy-plane, then

$$\iint_R f[x, y]\, dx\, dy > 0.$$

G.1.c.ii) Explain the statement:

If for a given function $f[x, y]$ and a given region R in the xy-plane, it turns out that

$$\iint_R f[x, y]\, dx\, dy < 0,$$

then you are guaranteed that there are some points $\{x, y\}$ in R with $f[x, y] < 0$.

■ G.2) Calculating double integrals*

G.2.a) Here's a region R plotted in the xy-plane.

```
In[3]:=
  Clear[y]
  Rplot = ParametricPlot[{{y(y - 6) + 1,y},
  {Cos[Pi y/6]^2,y}},{y,0,6},
  PlotStyle->{{Red,Thickness[0.01]}},
  AxesLabel->{"x","y"},
  Epilog->Text["R",{-4,3}]];
```

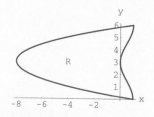

Go with

$$f[x, y] = y - x + 3$$

and calculate $\iint_R f[x, y]\, dx\, dy$ by the method of least labor on your part.

G.2.b) Here's another region R plotted in the xy-plane.

```
In[4]:=
  Clear[x,y,t]
  x[t_] = Cos[t](1 - Cos[t]);
  y[t_] = Sin[t](1 - Cos[t]);
  Rplot = ParametricPlot[{x[t],y[t]},{t,0,2 Pi},
  PlotRange->All,PlotStyle->{{Red,Thickness[0.01]}},
  AxesLabel->{"x","y"},Epilog->Text["R",{-1,0.5}]];
```

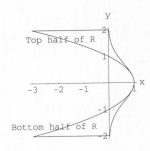

Go with

$$f[x,y] = 3 + y - x$$

and calculate $\iint_R f[x,y]\,dx\,dy$ by the method of least labor on your part. Next, go with $g[x,y] = 3 - x$ and calculate $\iint_R g[x,y]\,dx\,dy$ by the method of least labor on your part.

Compare the values of

$$\iint_R f[x,y]\,dx\,dy \qquad \text{and} \qquad \iint_R g[x,y]\,dx\,dy.$$

Discuss how you think that the shape and position of R account for the relationship between these two numbers.

G.2.c.i) Here is another region R plotted in the xy-plane.

```
In[5]:=
  Clear[high,low,y]; xhigh[y_] = E^(-y^2);
  xlow[y_] = 1 - y^2; Rplot = ParametricPlot[{
  {xhigh[y],y},{xlow[y],y}},{y,-2,2},PlotStyle->Red,
  AxesLabel->{"x","y"},PlotRange->All,AspectRatio->
  Automatic,Epilog->{{Red,Line[{{xlow[-2],-2},
  {xhigh[-2],-2}}]},{Red,Line[
  {{xlow[2],2},{xhigh[2],2}}]}},
  Text["Top half of R",{-3,1.7}],
  Text["Bottom half of R",{-3,-1.7}]}];
```

Go with

$$f[x,y] = y\,e^{-x},$$
$$g[x,y] = y^2\,e^{-x},$$

and

$$h[x,y] = y^3\,e^{-x}.$$

Calculate

$$\iint_R f[x,y]\,dx\,dy, \iint_R g[x,y]\,dx\,dy \qquad \text{and} \qquad \iint_R h[x,y]\,dx\,dy$$

by the method of least labor on your part.

G.2.c.ii)

Look at the formulas for $f[x, y]$ and $g[x, y]$ and then look at the shape and position of R and explain why it was no surprise that

$$\iint_R f[x, y]\, dx\, dy \qquad \text{and} \qquad \iint_R h[x, y]\, dx\, dy$$

came out the way they did.

G.2.d) Here is another region R plotted in the xy-plane.

```
In[6]:=
Clear[x,y,Rplot]
Rplot = Plot[{3 Sin[(Pi/4)x]^2,
x(4 - x)},{x,0,4},PlotStyle->
{{Red,Thickness[0.01]}},
AxesLabel->{"x","y"},
Epilog->Text["R",{2,3.5}]];
```

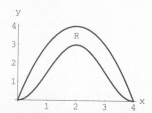

Go with $f[x, y] = 4\, x - y$ and calculate

$$\iint_R f[x, y]\, dx\, dy$$

by the method of least labor on your part.

G.2.e)

Given that the region R consists of everything inside the ellipse

$$\left(\frac{x + 3}{7}\right)^2 + \left(\frac{y - 1}{2}\right)^2 = 1,$$

go with $f[x, y] = x^2 - 3\, y^2$ and calculate

$$\iint_R f[x, y]\, dx\, dy$$

by the method of least labor on your part.

■ G.3) Area and volume measurements via Gauss-Green*

This problem has lots of overlap with some of the work in the lesson on transformations of integrals.

The Gauss-Green formula says

$$\iint_R D[n[x, y], x] - D[m[x, y], y]\, dx\, dy = \int_a^b m[x[t], y[t]]\, x'[t] + n[x[t], y[t]]\, y'[t]\, dt$$

provided that $\{x[t], y[t]\}$ sweeps out the boundary of the region R exactly one time in the counterclockwise fashion as t advances from a to b.

If you take $m[x, y] = y$ and $n[x, y] = 0$ and plug into the Gauss-Green formula, then you get

$$-\iint_R 1 \, dx \, dy = \iint_R (0 - 1) \, dx \, dy$$

$$= \int_a^b (y[t] \, x'[t] + 0 \, y'[t]) \, dt$$

$$= \int_a^b y[t] \, x'[t] \, dt.$$

Consequently, $-\int_a^b y[t] \, x'[t] \, dt$ measures the area of R provided that $\{x[t], y[t]\}$ sweeps out the boundary of R exactly one time in the counterclockwise fashion as t advances from a to b.

G.3.a.i) Adapt the discussion above to explain why

$$\int_a^b x[t] \, y'[t] \, dt$$

also measures the area of R provided that $\{x[t], y[t]\}$ sweeps out the boundary of R exactly one time in the counterclockwise fashion as t advances from a to b.

G.3.a.ii) Explain why

$$\left(\frac{1}{2}\right) \int_a^b (x[t] \, y'[t] - y[t] \, x'[t]) \, dt$$

measures the area of R provided that $\{x[t], y[t]\}$ sweeps out the boundary of R exactly one time in the counterclockwise fashion as t advances from a to b.

G.3.a.iii) Here's the plot of a whimsical region R in the xy-plane.

```
In[7]:=
Clear[x,y,t]
x[t_] = Cos[t](1 - Sin[t] - Cos[t]);
y[t_] = Sin[t](3 - Cos[t] - Sin[t]);
Rplot = ParametricPlot[{x[t],y[t]},{t,0,2 Pi},
PlotRange->All,PlotStyle->{{Red,Thickness[0.01]}},
AxesLabel->{"x","y"},Epilog->Text["R",{-1,1/2}]];
```

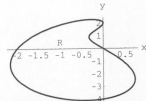

Choose any of the three formulas explained above to measure the area of this region.

G.3.a.iv) Here's the plot of another region R in the xy-plane.

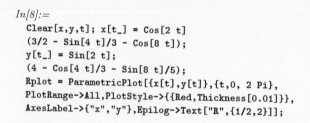
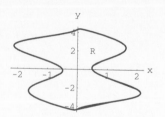

```
In[8]:=
  Clear[x,y,t]; x[t_] = Cos[2 t]
  (3/2 - Sin[4 t]/3 - Cos[8 t]);
  y[t_] = Sin[2 t];
  (4 - Cos[4 t]/3 - Sin[8 t]/5);
  Rplot = ParametricPlot[{x[t],y[t]},{t,0, 2 Pi},
  PlotRange->All,PlotStyle->{{Red,Thickness[0.01]}},
  AxesLabel->{"x","y"},Epilog->Text["R",{1/2,2}]];
```

Using the formula

$$\int_a^b x[t]\, y'[t]\, dt \qquad \text{with } a = 0 \text{ and } b = 2\pi$$

to measure the area inside this region, you get the area measurement:

```
In[9]:=
  Integrate[x[t] y'[t],{t,0,2 Pi}]
```

```
Out[9]=
  129 Pi
  ─────
   10
```

Unfortunately, this measurement is wrong.

> Find the error and come up with the correct measurement.

G.3.b) Here's a plot of the surface

$$z = f[x, y] = 5 - 0.2\left(x^2 + y^2\right)$$

above the region R in the xy-plane consisting of all points $\{x, y, 0\}$ with $\{x, y\}$ inside the ellipse

$$\left(\frac{x+1}{2}\right)^2 + \left(\frac{y+1}{3}\right)^2 = 1:$$

```
In[10]:=
  Clear[f,x,y,r,t]; f[x_,y_] = 5 - 0.2 (x^2 + y^2);
  {x[r_,t_],y[r_,t_]} = {-1,-1} + r {2 Cos[t],3 Sin[t]};
  surfaceplot = ParametricPlot3D[Evaluate[{x[r,t],y[r,t],f[x[r,t],
  y[r,t]]}],{t,0,2 Pi},{r,0,1},DisplayFunction->Identity];
  Rplot = ParametricPlot3D[Evaluate[{x[r,t],y[r,t],0}],{t,0,2 Pi},{r,0,1},
  PlotPoints->{Automatic,2},DisplayFunction->Identity];
  label = Graphics3D[Text["R",{-1,1,0}]]; spacer = 0.2;
  threedims = Graphics3D[{{Blue,Line[{{-3,0,0},{1.5,0,0}}]},
  Text["x",{1.5 + spacer,0,0}],{Blue,Line[{{0,-4,0},{0,2.5,0}}]},
  Text["y",{0,2.5 + spacer,0}],{Blue,Line[{{0,0,0},{0,0,5.5}}]},
  Text["z",{0,0,5.5 + spacer}]}];
```

In[11]:=
```
CMView = {2.7,1.6,1.2};
Show[surfaceplot,Rplot,label,threedims,
ViewPoint->CMView,PlotRange->All,
DisplayFunction->$DisplayFunction];
```

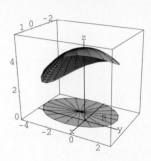

One way of measuring the volume of the solid consisting of all points between or on the ellipse and the surface plotted above is to realize that the top surface is just a plot of $f[x, y]$ for points $\{x, y, 0\}$ inside the ellipse R in the xy-plane. The volume measurement is just $\iint_R f[x, y] \, dx \, dy$, which you can easily calculate with the Gauss-Green formula once you remember the ellipse

$$\left(\frac{x+1}{2}\right)^2 + \left(\frac{y+1}{3}\right)^2 = 1$$

(which is the boundary of R) is parameterized in the counterclockwise way by

$$\{x[t], y[t]\} = \{-1, -1\} + \{2 \cos[t], 3 \sin[t]\} \qquad \text{with } 0 \le t \le 2\pi.$$

Now you can turn the winning team of *Mathematica* and the Gauss-Green formula loose:

In[12]:=
```
Clear[n,m,x,y,s,t]; m[x_,y_] = 0;
n[x_,y_] = Integrate[f[s,y],{s,0,x}];
a = 0; b = 2 Pi; {x[t_],y[t_]} = {-1,-1} + {2 Cos[t],3 Sin[t]};
volume = NIntegrate[m[x[t],y[t]] x'[t] + n[x[t],y[t]] y'[t],{t,a,b}]
```

Out[12]=
74.4557

Now look at this alternate calculation:

In[13]:=
```
Clear[n,m,x,y,s,t]; n[x_,y_] = 0;
m[x_,y_] = Integrate[-f[x,s],{s,0,y}];
a = 0; b = 2 Pi; {x[t_],y[t_]} = {-1,-1} + {2 Cos[t],3 Sin[t]};
volume = NIntegrate[m[x[t],y[t]] x'[t] + n[x[t],y[t]] y'[t],{t,a,b}]
```

Out[13]=
74.4557

Use the Gauss-Green formula

$$\iint_R D[n[x, y], x] - D[m[x, y], y] \, dx \, dy$$

$$= \int_a^b m[x[t], y[t]] \, x'[t] + n[x[t], y[t]] \, y'[t] \, dt$$

and the fundamental formula of calculus to explain why the alternate calculation was guaranteed to work.

■ G.4) Plot3D versus ParametricPlot3D⋆

G.4.a.i) Go with $f[x] = x^2\, e^{-x}$ and look at:

```
In[14]:=
  Clear[f,x]
  {a,b} = {-0.6,8};
  f[x_] = x^2 E^(-x);
  raw = Plot[f[x],
  {x,a,b},PlotRange->All];
```

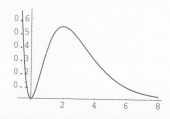

Explain with words and pictures what you think this plot represents.

G.4.a.ii) Stay with the same function $f[x]$ and the same values for a and b and look at:

```
In[15]:=
  parametric =
  ParametricPlot[
  {x,f[x]},{x,a,b},
  PlotRange->All];
```

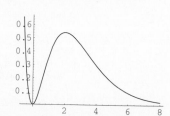

Say what happened, and try to explain why it will happen no matter what function $f[x]$ you go with.

G.4.b) Go with

$$f[x,y] = 2 + \sin[y] - x$$

and look at:

```
In[16]:=
  Clear[f,x,y]
  {{a,b},{c,d}} = {{-2,2},{-Pi/3,Pi/2}};
  f[x_,y_] = 2 + Sin[2 y] + x/2;
  CMView = {2.7,1.6,1.2};
  raw = Plot3D[f[x,y],{x,a,b},{y,c,d},
  ViewPoint->CMView,BoxRatios->Automatic];
```

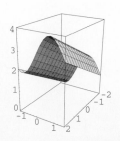

Explain with words and pictures what you think this plot represents.

G.4.c) Look at this plot:

```
In[17]:=
  Clear[f,x,y]
  {a,b} = {3,2}; f[x_,y_] = Sin[x - y];
  regular = Plot3D[f[x,y],{x,-a,a},{y,-b,b},
  ViewPoint->CMView,BoxRatios->Automatic];
```

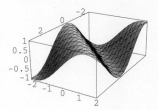

Stay with the same function $f[x,y]$ and the same a and b, and look at this parametric plot:

```
In[18]:=
  parametric = ParametricPlot3D[
  {x,y,f[x,y]},{x,-a,a},{y,-b,b},
  ViewPoint->CMView,BoxRatios->Automatic];
```

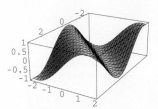

Say what happened, and try to explain why it will happen no matter what function $f[x,y]$ you go with.

■ G.5) Big league plotting and measuring

This problem appears only in the electronic version.

■ G.6) Average value and centroids

Most math folks say that the average value of a function $f[x]$ on an interval $[a, b]$ is the number A satisfying

$$\int_a^b f[x]\, dx = A\,(b - a) = A \int_a^b dx.$$

Going with this formula, you see that the average value of $\cos[x]$ on $[1, 3]$ is given by:

```
In[19]:=
  A = NIntegrate[Cos[x],{x,1,3}]/(3 - 1)
```

```
Out[19]=
  -0.350175
```

The following routine calculates cos[x] for 50 random points x chosen out of $[1,3]$ by *Mathematica*:

In[20]:=
```
Clear[k]
randoms = Table[N[Cos[Random[Real,{1,3}]]],{k,1,50}]
```
Out[20]=
```
{-0.675527, -0.567998, -0.952493, -0.122416, -0.442938, 0.207961, -0.956569, 0.435183,
 -0.96361, -0.706358, -0.710218, 0.271707, 0.445179, -0.950886, -0.716643, 0.250431,
 -0.20879, -0.789947, -0.750962, -0.504695, 0.51414, -0.475706, -0.941738, 0.488015,
 -0.950604, -0.516793, -0.684047, -0.534399, -0.41848, -0.412598, 0.0945104,
 -0.987367, -0.951817, -0.680462, -0.335019, -0.086822, 0.159489, -0.867522,
 0.0420343, -0.969425, -0.771222, -0.00651292, 0.373564, -0.784488,
 0.404731, -0.644632, 0.364627, -0.0342516, 0.464557, 0.456169}
```

Average these values:

In[21]:=
```
average = Sum[randoms[[k]], {k,1,Length[randoms]}]/Length[randoms]
```
Out[21]=
```
-0.342033
```

Do it again for 100 randomly chosen points from $[1,3]$:

In[22]:=
```
Clear[k]
randoms = Table[N[Cos[Random[Real,{1,3}]]],{k,1,100}];
average = Sum[randoms[[k]],{k,1,Length[randoms]}]/Length[randoms]
```
Out[22]=
```
-0.359673
```

Do it again for 300 randomly chosen points from $[1,3]$:

In[23]:=
```
randoms = Table[N[Cos[Random[Real,{1,3}]]],{k,1,300}];
average = Sum[randoms[[k]],{k,1,Length[randoms]}]/Length[randoms]
```
Out[23]=
```
-0.347587
```

G.6.a) In your opinion, how is the average value of cos[x] on $[1,3]$ calculated through the formula

$$\frac{1}{3-1}\int_1^3 \cos[x]\,dx,$$

as above, related to the outcomes of the numerical experiments?

G.6.b.i) Calculate the average value of sin[x] on $[0,\pi]$ in decimals.

G.6.b.ii) Look at:

In[24]:=
```
randoms = Table[N[Sin[Random[Real,{0,N[Pi]}]]],{k,1,300}];
average = Sum[randoms[[k]],{k,1,Length[randoms]}]/Length[randoms]
```

Out[24]=
```
0.610855
```

In[25]:=
```
randoms = Table[N[Sin[Random[Real,{0,N[Pi]}]]],{k,1,300}];
average = Sum[randoms[[k]],{k,1,Length[randoms]}]/Length[randoms]
```

Out[25]=
```
0.648061
```

In[26]:=
```
randoms = Table[N[Sin[Random[Real,{0,N[Pi]}]]],{k,1,300}];
average = Sum[randoms[[k]],{k,1,Length[randoms]}]/Length[randoms]
```

Out[26]=
```
0.641878
```

In[27]:=
```
randoms = Table[N[Sin[Random[Real,{0,N[Pi]}]]],{k,1,300}];
average = Sum[randoms[[k]],{k,1,Length[randoms]}]/Length[randoms]
```

Out[27]=
```
0.618923
```

Discuss the results.

G.6.c) You saw above that the average value of a function $f[x]$ on an interval $[a, b]$ on the x-axis is the number A satisfying

$$\int_a^b f[x]\,dx = A\,(b - a) = A \int_a^b dx.$$

Math folks have a similar notion for functions of two variables; they say: The average value of a function $f[x, y]$ on a region R in the xy-plane is the number A satisfying

$$\iint_R f[x, y]\,dx\,dy = A \iint_R dx\,dy.$$

Calculate the average value of

$$f[x, y] = x^2 + y^2$$

on the rectangle R consisting of the points with $-3 \leq x \leq 3$ and $-2 \leq y \leq 2$.

G.6.d.i) The centroid of a region R in the xy-plane is the point $\{x^*, y^*\}$ where

x^* is the average value of $f[x, y] = x$ on R

and

y^* is the average value of $g[x, y] = y$ on R.

> Calculate the centroid of the rectangle R consisting of the points with $-3 \leq x \leq 3$ and $-2 \leq y \leq 2$. Are you surprised by the result?

G.6.d.ii) Here's region R plotted in the xy-plane:

```
In[28]:=
Clear[x,y,t]
x[t_] = Cos[t](1 - Cos[t]);
y[t_] = Sin[t](1 - Cos[t]);
Rplot = ParametricPlot[
{x[t],y[t]},{t,0,2 Pi},
PlotRange->All,PlotStyle->
{{Red,Thickness[0.01]}},
AxesLabel->{"x","y"},
Epilog->Text["R",{-1,0.75}]];
```

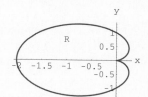

Fancy dudes from Ivy League schools say that this is the cardioid described in polar coordinates by the polar equation $r[t] = 1 - \cos[t]$.

> Use the Gauss-Green formula to help you calculate the centroid of this region.

G.6.d.iii) Here's another region R plotted in the xy-plane:

```
In[29]:=
Clear[x,Rplot]
Rplot = Plot[{3 Sin[(Pi/4)x]^2,
x(4 - x)},{x,0,4},
PlotStyle->{{Red,Thickness[0.01]}},
AxesLabel->{"x","y"},
Epilog->Text["R",{2,3}]];
```

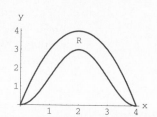

> Calculate the centroid of this region. Is the centroid inside or outside the region?

G.6.d.iv) Sometimes you can look at a plot of a region R and know before you do any calculation that the centroid of R will calculate out to be inside R.

> What is it about the bulges and hollows of the plot of the boundary of R that helps you to do this?

■ G.7) Bombing runs

G.7.a.i) You've got a target sitting at $\{0,0\}$ and you want to knock it out with an iron bomb. If you set the Northrup bomb sight to hit the target at $\{0,0\}$ and examine the point $\{x,y\}$ at which an individual bomb lands, you will find that the resulting coordinates x and y

→ have no influence on each other;

→ are both normally distributed; and

→ both average out to $\{0,0\}$.

You will also find that the standard deviation of the x measurement is not the same as the standard deviation of the y measurement. For a specific bomb and airplane, it turns out that

   xdev = 50 feet

and

   ydev = 100 feet.

And to destroy the target, this type of bomb must hit within 100 feet of the target.

> Estimate the fraction of the bombs of this type, aimed at this target, that actually destroy the target.

G.7.a.ii)

> In part G.7.a.i) above, what are your chances of destroying the target if you bomb it twice?
>
> How many times do you have to bomb the target to get 95% confidence of destroying it?

G.7.b.i) This time your target is a long, thin radar installation 100 feet long centered on $\{0,0\}$ and visualized as the line segment running from $\{-50,0\}$ to $\{50,0\}$:

```
In[30]:=
  radarstation = Show[
  Graphics[{Blue,Thickness[0.02],
  Line[{{-50,0},{50,0}}]}],
  Graphics[Text["Radar Station",{0,5}]],
  PlotRange->{{-80,80},{-30,30}},
  Axes->True,AxesLabel->{"x","y"}];
```

The Air Force bomb damage assessment analysts have determined that to knock out this radar station, you must place your iron bomb within the ellipse

$$\left(\frac{x}{60}\right)^2 + \left(\frac{y}{100}\right)^2 = 1 :$$

In[31]:=
```
Clear[x,y,t]; x[t_] = 60 Cos[t]; y[t_] = 100 Sin[t];
ellipse = ParametricPlot[{x[t],y[t]},{t,0,2 Pi},
PlotStyle->{{Thickness[0.01],Red}},
DisplayFunction->Identity];
Show[ellipse,radarstation,
AspectRatio->Automatic,PlotRange->All,
DisplayFunction->$DisplayFunction];
```

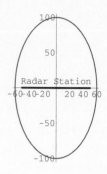

Here are two reasons for why the shape is an ellipse:

→ The long, flat nature of the radar station makes it very vulnerable to bombs that land to its front or rear.

→ The same long, flat nature of the radar station makes it almost invulnerable to bombs that land to its left or right.

The bombs you are going to use have

$$\text{xdev} = 40 \text{ feet} \qquad \text{and} \qquad \text{ydev} = 80 \text{ feet}.$$

Estimate the fraction of the bombs aimed at $\{0,0\}$ that actually destroy the radar station.

G.7.b.ii) In part G.7.b.i) above, what are your chances of destroying the radar station if you bomb it twice aiming the bombs at $\{0,0\}$?

How many times do you have to bomb the radar station, aiming the bombs at $\{0,0\}$, to get 95% confidence of destroying the radar station?

LESSON 2.06

More Tools and Measurements

Basics

■ **B.1)** **Separating the variables and integrating to solve some differential equations**

B.1.a) Here is *Mathematica*'s formula for the solution of the differential equation

$$y'[x] = x\,(y[x] + 4)^2 \qquad \text{with } y[3] = 9.$$

In[1]:=
```
Clear[y,x,sol]
sol=DSolve[{y'[x] == x (y[x] + 4)^2,y[3] == 9},y[x],x]
```

Out[1]=

$$\{\{y[x] \,\text{->}\, \frac{2\,(225\,-\,26\,x^2)}{-119\,+\,13\,x^2}\}\}$$

> Use the technique of separating and integrating to explain where this formula comes from.

Answer: Rewrite $y'[x] = x\,(y[x] + 4)^2$ and get

$$\frac{y'[x]}{(y[x] + 4)^2} = x.$$

Some folks call this technique of putting all the $y[x]$ terms on one side and the x terms on the other side by the name "separation of variables." Integrate both sides from 3 to x to get

$$\int_3^x \frac{y'[t]}{(y[t]+4)^2}\,dt = \int_3^x t\,dt,$$

remembering that $y[3] = 9$:

In[2]:=
```
Clear[x,y,t]
left = Integrate[y'[t]/(y[t] + 4)^2,{t,3,x}]/.y[3]->9
```

Out[2]=

$$\frac{1}{13} - \frac{1}{4 + y[x]}$$

In[3]:=
```
right = Integrate[t,{t,3,x}]
```

Out[3]=

$$-\left(\frac{9}{2}\right) + \frac{x^2}{2}$$

Here comes the formula for $y[x]$:

In[4]:=
```
Solve[left == right,y[x]]
```

Out[4]=

$$\{\{y[x] \;\rightarrow\; -\left(\frac{450 - 52\ x^2}{119 - 13\ x^2}\right)\}\}$$

Compare:

In[5]:=
```
sol
```

Out[5]=

$$\{\{y[x] \;\rightarrow\; \frac{2\ (225 - 26\ x^2)}{-119 + 13\ x^2}\}\}$$

Same thing. Not bad, eh?

B.1.b)   Here is *Mathematica*'s formula for the solution of the logistic differential equation

$$y'[x] = a\,y[x]\left(\frac{1 - y[x]}{b}\right) \qquad \text{with } y[0] = c.$$

In[6]:=
```
Clear[y,x,a,b,c,sol]
sol = DSolve[{y'[x] == a y[x] (1 - y[x]/b),y[0] == c},y[x],x]
```

Out[6]=

$$\{\{y[x] \;\rightarrow\; \frac{b\ c\ E^{a\ x}}{b - c + c\ E^{a\ x}}\}\}$$

Use the technique of separating and integrating to explain where this formula comes from.

Answer: Rewrite $y'[x] = a\,y[x]\,(1 - y[x]/b)$ by separating the variables to get:

$$\frac{y'[x]}{y[x]\,(1 - y[x]/b)} = a.$$

Integrate both sides from 0 to x to get

$$\int_0^x \frac{y'[t]}{y[t]\,(1 - y[t]/b)}\,dt = \int_0^x a\,dt,$$

remembering that $y[0] = c$:

In[7]:=
```
Clear[x,y,t]
left = Integrate[y'[t]/(y[t](1 - y[t]/b)),{t,0,x}]/.y[0]->c
```

Out[7]=
```
Log[b - c] - Log[c] - Log[b - y[x]] + Log[y[x]]
```

Put this in better form, remembering that

$$\log[p] - \log[q] - \log[r] + \log[s] = \log\left[\frac{p\,s}{q\,r}\right].$$

In[8]:=
```
Clear[p,q,r,s];
logrules = {Log[p_] + Log[q_]->Log[p q],-Log[r_]->Log[1/r]};
betterleft = left//.logrules
```

Out[8]=
```
     (b - c) y[x]
Log[-------------]
      c (b - y[x])
```

In[9]:=
```
right = Integrate[a,{t,0,x}]
```

Out[9]=
```
a x
```

Exponentiate both sides:

In[10]:=
```
newleft = E^betterleft
```

Out[10]=
```
 (b - c) y[x]
--------------
 c (b - y[x])
```

In[11]:=
```
newright = E^right
```

Out[11]=
```
 a x
E
```

Here comes the formula for $y[x]$:

In[12]:=
```
Solve[newleft == newright,y[x]]
```

Out[12]=

$$\{\{y[x] \; \rightarrow \; \frac{b \; c \; E^{a \; x}}{b \; - \; c \; + \; c \; E^{a \; x}}\}\}$$

Compare:

In[13]:=
```
sol
```

Out[13]=

$$\{\{y[x] \; \rightarrow \; \frac{b \; c \; E^{a \; x}}{b \; - \; c \; + \; c \; E^{a \; x}}\}\}$$

Same thing. Now you know where the formula for a logistic function comes from.

B.1.c) In situations in which it works, separation of variables is a powerful technique for producing formulas for solutions of differential equations.

> When is it possible to do this?

Answer: That's hard to say. If you've got something like

$$y'[x] = \sin[x] \, (y[x] - 6)^2 \qquad \text{with } y[0] = 5,$$

then you can separate the variables by writing it as

$$\frac{y'[x]}{(y[x] - 6)^2} = \sin[x]$$

and then try to integrate. If you've got something like

$$y'[x] = \sin[x] \, (y[x] - 6)^2 + \cos[x] \qquad \text{with } y[0] = 5,$$

then there is no way to separate the variables. In this case, separation of variables fails miserably, and you can try DSolve or NDSolve.

■ B.2) Integration by parts

B.2.a)

> Use the product rule for differentiation to explain why
>
> $$\int_a^b u[x] \, v'[x] \, dx = u[x] \, v[x] \Big|_a^b - \int_a^b v[x] \, u'[x] \, dx.$$

This is called the integration by parts formula. This formula can be especially handy for hand calculation of integrals.

Answer: The product rule says

$$\frac{d\left(u[x]\,v[x]\right)}{dx} = u[x]\,v'[x] + u'[x]\,v[x].$$

Integrating this from $x = a$ to $x = b$ gives:

$$\int_a^b \frac{d\left(u[x]\,v[x]\right)}{dx}\,dx = \int_a^b u[x]\,v'[x]\,dx + \int_a^b u'[x]\,v[x]\,dx.$$

This is the same as:

$$u[x]\,v[x]\Big|_a^b = \int_a^b u[x]\,v'[x]\,dx + \int_a^b u'[x]\,v[x]\,dx.$$

Rearranging gives

$$\int_a^b u[x]\,v'[x]\,dx = u[x]\,v[x]\Big|_a^b - \int_a^b v[x]\,u'[x]\,dx,$$

and this explains why the advertised formula holds up.

B.2.b.i)

Use the integration by parts formula to calculate

$$\int_0^1 x\,e^x\,dx.$$

Answer: The integration by parts formula is

$$\int_0^1 u[x]\,v'[x]\,dx = u[x]\,v[x]\Big|_0^1 - \int_0^1 v[x]\,u'[x]\,dx.$$

Make the assignments:

$$u[x] = x$$

and

$$v'[x] = e^x.$$

This gives $u'[x] = 1$ and $v[x] = e^x$. With these assignments,

$$\int_0^1 x\,e^x\,dx = \int_0^1 u[x]\,v'[x]\,dx$$

$$= u[x]\,v[x]\Big|_0^1 - \int_0^1 v[x]\,u'[x]\,dx$$

$$= x\,e^x\Big|_0^1 - \int_0^1 e^x\,1\,dx$$

$$= x\, e^x \Big[_0^1 - \int_0^1 e^x\, dx$$

$$= x\, e^x \Big[_0^1 - e^x \Big[_0^1$$

$$= (e - 0) - (e - 1) = 1.$$

Check:

In[14]:=
   NIntegrate[x E^x,{x,0,1}]

Out[14]=
   1.

Got it.

B.2.b.ii) The integration by parts formula is

$$\int_a^b u[x]\, v'[x] dx = u[x]\, v[x] \Big[_a^b - \int_a^b v[x]\, u'[x]\, dx.$$

When this formula was used to help calculate $\int_0^1 x\, e^x dx$ above, the assignments $u[x] = x$ and $v'[x] = e^x$ were made.

> Do you get into deep do-do if you make the assignments
>
> $$u[x] = e^x$$
>
> and
>
> $$v'[x] = x?$$

Answer: Try it and see. The integration by parts formula is

$$\int_0^1 u[x]\, v'[x]\, dx = u[x]\, v[x] \Big[_0^1 - \int_0^1 v[x]\, u'[x]\, dx.$$

Make the assignments:

$$u[x] = e^x$$

and

$$v'[x] = x.$$

This gives $u'[x] = e^x$ and $v[x] = x^2/2$. With these assignments,

$$\int_0^1 x\, e^x\, dx = \int_0^1 u[x]\, v'[x]\, dx$$

$$= u[x]\, v[x] \Big[_0^1 - \int_0^1 v[x]\, u'[x]\, dx$$

$$= e^x\, \frac{x^2}{2} \Big[_0^1 - \int_0^1 \frac{x^2}{2} e^x\, dx.$$

This puts you in deep do-do because now you have to calculate

$$\int_0^1 \left(\frac{x^2}{2}\right) e^x \, dx,$$

and this integral is harder to calculate than the original integral $\int_0^1 x e^x \, dx$.

B.2.b.iii) | What moral is suggested by part B.2.b.ii)?

Answer: When you use the integration by parts formula

$$\int_a^b u[x] \, v'[x] \, dx = u[x] \, v[x] \Big|_a^b - \int_a^b v[x] \, u'[x] \, dx,$$

you are free to assign $u[x]$ and $v'[x]$. Generally, you will want to make the assignments so that the integral on the right-hand side is easier to calculate than the integral on the left.

B.2.c.i) | Use the integration by parts formula to help to calculate
$$\int_0^\pi x \cos[x] \, dx.$$

Answer: The integration by parts formula is

$$\int_0^\pi u[x] \, v'[x] \, dx = u[x] \, v[x] \Big|_0^\pi - \int_0^\pi v[x] \, u'[x] \, dx.$$

You have your choice of the assignments:

$\rightarrow$ Choice 1: $u[x] = \cos[x]$ and $v'[x] = x$. This gives

$$u'[x] = \sin[x]$$

and

$$v[x] = \frac{x^2}{2},$$

and

$$v[x] \, u'[x] = \left(\frac{x^2}{2}\right) \sin[x].$$

$\rightarrow$ Choice 2: $u[x] = x$ and $v'[x] = \cos[x]$. This gives

$$u'[x] = 1$$

and

$$v[x] = \sin[x]$$

and

$$v[x] \, u'[x] = \sin[x].$$

Because $v[x] \, u'[x]$ is simpler with Choice 2 than it is with Choice 1, Choice 2 is the way to go. Plug Choice 2 into the integration by parts formula to get

$$
\int_0^\pi x \, \cos[x] \, dx = \int_0^\pi u[x] \, v'[x] \, dx
$$

$$
= u[x] \, v[x] \Big[_0^\pi - \int_0^\pi v[x] \, u'[x] \, dx
$$

$$
= x \, \sin[x] \Big[_0^\pi - \int_0^\pi \sin[x] \, dx
$$

$$
= x \, \sin[x] \Big[_0^\pi - (-\cos[x]) \Big[_0^\pi
$$

$$
= x \, \sin[x] \Big[_0^\pi + (\cos[x]) \Big[_0^\pi
$$

$$
= (0 - 0) + (-1 - 1) = -2.
$$

Check:

In[15]:=
```
NIntegrate[x Cos[x],{x,0,Pi}]
```
Out[15]=
```
-2.
```

Nailed it.

B.2.c.ii)   Use the integration by parts formula to help calculate

$$
\int_1^t \log[x] \, dx.
$$

Answer: The integration by parts formula is

$$
\int_0^\pi u[x] \, v'[x] \, dx = u[x] \, v[x] \Big[_0^\pi - \int_0^\pi v[x] \, u'[x] \, dx.
$$

Rewrite the integral as $\int_1^t \log[x] \, dx = \int_1^t \log[x] \, 1 \, dx$. You have your choice of the assignments:

$\rightarrow$ Choice 1: $u[x] = \log[x]$ and $v'[x] = 1$. This gives

$$
u'[x] = \frac{1}{x}
$$

and

$$
v[x] = x
$$

and

$$
v[x] \, u'[x] = x \left(\frac{1}{x} \right) = 1.
$$

$\rightarrow$ Choice 2: $u[x] = 1$ and $v'[x] = \log[x]$. This gives

$$
u'[x] = 0
$$

and

$$
v[x] = \text{(Choke and punt)}.
$$

Choice 2 stinks because it's hard to come up with $v[x]$. In fact, finding $v[x]$ is where we started. So Choice 1 is the way to go. This gives $u'[x] = 1/x$ and $v[x] = x$. Plug Choice 1 into the integration by parts formula to get

$$\int_1^t \log[x]\,1\,dx = \int_1^t u[x]\,v'[x]\,dx$$

$$= u[x]\,v[x]\big[_1^t - \int_1^t v[x]\,u'[x]\,dx$$

$$= \log[x]\,x\big[_1^t - \int_1^t x\left(\frac{1}{x}\right)\,dx$$

$$= \log[x]\,x\big[_1^t - \int_1^t 1\,dx$$

$$= \log[x]\,x\big[_1^t - x\big[_1^t$$

$$= (t\,\log[t] - 0) - (t - 1) = t\,\log[t] - t + 1.$$

Check:

In[16]:=
```
Clear[x,t]; integral = Integrate[Log[x],{x,1,t}]
```

Out[16]=
```
1 + t (-1 + Log[t])
```

In[17]:=
```
Expand[integral]
```

Out[17]=
```
1 - t + t Log[t]
```

Nailed it cold.

■ B.3)   Complex numbers, the formula $e^{a+ib} = e^a\,(\cos[b] + i\,\sin[b])$, and the logarithm of a negative number

The imaginary number i enters the door to mathematics as a solution of $x^2 + 1 = 0$:

In[18]:=
```
Clear[x]; Solve[x^2 + 1 == 0]
```

Out[18]=
```
{{x -> I}, {x -> -I}}
```

Mathematica writes $i = \sqrt{-1}$ as a capital.

Check:

In[19]:=
```
I^2
```

Out[19]=
```
-1
```

In[20]:=
```
(-I)^2
```

Out[20]=
```
-1
```

Once $i = \sqrt{-1}$ comes in the door, then so do the complex numbers. A complex number is any number $a + ib$ where a and b are real numbers. Complex numbers come up as solutions of quadratic equations that have no real solutions:

In[21]:=
```
Clear[x]; Solve[x^2+x+1 == 0]
```

Out[21]=

$$\left\{\left\{x \to \frac{-1 + I\ \text{Sqrt}[3]}{2}\right\}, \left\{x \to \frac{-1 - I\ \text{Sqrt}[3]}{2}\right\}\right\}$$

This tells you that the complex numbers

$$-\frac{1}{2} + i\,\frac{\sqrt{3}}{2} \qquad \text{and} \qquad -\frac{1}{2} - i\,\frac{\sqrt{3}}{2}$$

both solve the equation $x^2 + x + 1 = 0$.

→ You can add complex numbers:

In[22]:=
```
(1 + I) + (1 + 3 I)
```

Out[22]=
```
2 + 4 I
```

In[23]:=
```
(2 + 2 I) + (-2 + 2 I)
```

Out[23]=
```
4 I
```

→ You can subtract complex numbers:

In[24]:=
```
(3 + 4 I) - (2 + 3 I)
```

Out[24]=
```
1 + I
```

In[25]:=
```
(1 + I) - (1 + 3 I)
```

Out[25]=
```
-2 I
```

→ You can multiply complex numbers:

In[26]:=
```
I^2
```

Out[26]=
```
-1
```

In[27]:=
```
(2 + 3 I) I
```
Out[27]=
```
-3 + 2 I
```

In[28]:=
```
(1 + I)^4
```
Out[28]=
```
-4
```

In[29]:=
```
Expand[(1 + Sqrt[3] I)^3]
```
Out[29]=
```
-8
```

In[30]:=
```
Expand[(1 + I)^2]
```
Out[30]=
```
2 I
```

$\rightarrow$ You can divide complex numbers:

In[31]:=
```
(1 + 2 I)/I
```
Out[31]=
```
2 - I
```

In[32]:=
```
(3 + 2 I)/(2 - 6 I)
```
Out[32]=
```
     3       11 I
  -(---)  +  ----
    20        20
```

B.3.a) Here is *Mathematica*'s calculation of

$$f[t] = \int_{-5}^{t} \frac{1}{x}\, dx :$$

In[33]:=
```
Clear[f,x,t]
f[t_] = Integrate[1/x,{x,-5,t}]
```
Out[33]=
```
-I Pi - Log[5] + Log[t]
```

Take another look at this integral $f[t] = \int_{-5}^{t} 1/x\, dx$, remembering that you can plug in only negative t's because you cannot integrate through the singularity of $1/x$ at $x = 0$.

In[34]:=
```
Clear[f,x,t]
f[t_] = Integrate[1/x,{x,-5,t}]
```

Out[34]=
```
 -I Pi - Log[5] + Log[t]
```

When you plug in $t = -2$, you get

$$f[-2] = -i\pi - \log[5] + \log[-2].$$

Mathematica is not bothered by this:

In[35]:=
```
 f[-2]
```

Out[35]=
```
 Log[2] - Log[5]
```

Mathematica made sense of $\log[-2]$. This might pose a difficulty for you because you have probably been told again and again that you can't take a logarithm of a negative number. *Mathematica* has not heard about this dogma; see what *Mathematica* thinks about the natural logarithm of -2:

In[36]:=
```
 Log[-2]
```

Out[36]=
```
 I Pi + Log[2]
```

Yikes!

$$\log[-2] = \log[2] + i\pi.$$

In[37]:=
```
 Sqrt[-1]
```

Out[37]=
```
 I
```

Do some additional prospecting:

In[38]:=
```
 Log[-4]
```

Out[38]=
```
 I Pi + Log[4]
```

$$\log[-4] = \log[4] + i\pi$$

In[39]:=
```
 Log[-3]
```

Out[39]=
```
 I Pi + Log[3]
```

The pattern holds. If you want the value of $\log[x]$ to be a real number for $x < 0$, then you can't do it. But if you are willing to resort to complex numbers, then you can do it rather easily. Here's how it goes: For a real number y,

$$y = \log[x]$$

means

$$e^y = x.$$

Because $e^y = x$ is always positive, this tells you that if you stick to real numbers, then you can take the natural logarithm of positive numbers only. On the other hand, if you have a meaning for

$$e^{a+ib} \qquad \text{for} \qquad i = \sqrt{-1},$$

you open the door to logarithms for negative numbers. The way folks all across our planet give meaning to e^{a+ib} is to agree that

$$e^{a+ib} = e^a \left(\cos[b] + i \sin[b] \right).$$

What's the basis for this agreement that

$$e^{a+ib} = e^a (\cos[b] + i \sin[b])?$$

Answer: To begin with, it should be true that

$$e^{a+ib} = e^a \, e^{ib}.$$

And you want

$$e^{a+ib} = e^a \left(\cos[b] + i \sin[b] \right).$$

There is no problem with this if you agree that

$$e^{ib} = \cos[b] + i \sin[b].$$

So the basis for understanding the agreement $e^{a+ib} = e^a \left(\cos[b] + i \sin[b] \right)$ boils down to understanding why the agreement

$$e^{ib} = \cos[b] + i \sin[b]$$

is a good deal for everyone. Check it out:

$\rightarrow \; e^{i0}$ should be $e^0 = 1$:

In[40]:=
```
Clear[b]
(Cos[b] + I Sin[b])/.b->0
```
Out[40]=
```
1
```

This checks.

$\rightarrow \; e^{ib} e^{ic}$ should be $e^{i(b+c)} = \cos[b+c] + i \sin[b+c]$:

In[41]:=
```
Clear[b,c]
Expand[(Cos[b] + I Sin[b]) (Cos[c] + I Sin[c]),Trig->True]
```
Out[41]=
```
Cos[b + c] + I Sin[b + c]
```

This checks.

$\rightarrow$ The derivative of e^{ib} with respect to b should be

$$i\,e^{ib} = i\,(\cos[b] + i\,\sin[b]) = i\,\cos[b] - \sin[b]\,:$$

In[42]:=
```
D[Cos[b] + I Sin[b],b]
```
Out[42]=
```
I Cos[b] - Sin[b]
```

This checks. Can you think of anything else to check before you agree that

$$e^{ib} = \cos[b] + i\,\sin[b]?$$

If you can, then check it out for yourself. You will be happy with the results. *Mathematica* joins in the agreement:

In[43]:=
```
ComplexExpand[E^(I b)]
```
Out[43]=
```
Cos[b] + I Sin[b]
```

In[44]:=
```
ComplexExpand[E^(a + I b)]
```
Out[44]=
```
 a          a
E  Cos[b] + I E  Sin[b]
```

The complex exponential $e^{a+ib} = e^a\,(\cos[b] + i\,\sin[b])$ is one of the few issues in which the whole world is in perfect harmony.

B.3.b.i) Look at:

In[45]:=
```
Log[-1]
```
Out[45]=
```
I Pi
```

How does the agreement $e^{a+ib} = e^a\,(\cos[b] + i\,\sin[b])$ explain why

$$\log[-1] = i\,\pi?$$

Answer: Saying $\log[-1] = i\,\pi$ is the same as saying $e^{i\pi} = \cos[\pi] + i\,\sin[\pi] = -1$.

Check it out:

In[46]:=
```
Clear[b]
(Cos[b] + I Sin[b])/.b->Pi
```
Out[46]=
```
-1
```

This checks. Or use the built-in *Mathematica* function:

In[47]:=
```
ComplexExpand[E^(I Pi)]
```
Out[47]=
```
-1
```

Good.

B.3.b.ii) Look at:

In[48]:=
```
Log[-2]
```
Out[48]=
```
I Pi + Log[2]
```

> How does the agreement $e^{a+ib} = e^a \left(\cos[b] + i \sin[b] \right)$ explain why
>
> $$\log[-2] = \log[2] + i\pi?$$

Answer: Saying $\log[-2] = \log[2] + i\pi$ is the same as saying

$$e^{\log[2]+i\pi} = e^{\log[2]} \left(\cos[\pi] + i \sin[\pi] \right) = -2.$$

Check it out:

In[49]:=
```
Clear[a,b]
(E^a)(Cos[b] + I Sin[b])/.{a->Log[2],b->Pi}
```
Out[49]=
```
-2
```

This checks. Or use the built-in *Mathematica* function:

In[50]:=
```
ComplexExpand[E^(Log[2] + I Pi)]
```
Out[50]=
```
-2
```

Nice.

B.3.b.iii)

> How does the agreement $e^{a+ib} = e^a \left(\cos[b] + i \sin[b] \right)$ explain why
>
> $$\log[-x] = \log[x] + i\pi$$
>
> for any positive (real) number x?

Answer: Saying $\log[-x] = \log[x] + i\pi$ is the same as saying

$$e^{\log[x]+i\pi} = e^{\log[x]} \left(\cos[\pi] + i \sin[\pi] \right) = -x.$$

Check it out:

In[51]:=
```
Clear[a,b,x]; (E^a)(Cos[b] + I Sin[b])/.{a->Log[x],b->Pi}
```

Out[51]=
```
-x
```

This checks. Or use the built-in *Mathematica* function:

In[52]:=
```
Expand[E^(Log[x] + I Pi),Trig->True]
```

Out[52]=
```
-x
```

Cool.

B.3.c)   Look at:

In[53]:=
```
Clear[f,x]; f[x_] = Log[x]; f'[x]
```

Out[53]=
$$\frac{1}{x}$$

In the calculation above, *Mathematica* doesn't know whether x is positive or negative:

In[54]:=
```
f'[3]
```

Out[54]=
$$\frac{1}{3}$$

In[55]:=
```
f'[-3]
```

Out[55]=
$$-\left(\frac{1}{3}\right)$$

Apparently the derivative of log[x] is $1/x$, no matter whether $x > 0$ or $x < 0$.

> Explain where this comes from.

Answer: Put $f[x] = \log[x]$. You already know that if x is positive, then $f'[x] = 1/x$.

On the other hand, if x is positive, then $-x$ is negative and $f[-x] = \log[x] + i\,\pi$. The chain rule tells you that $D[f[-x], x] = -f'[-x]$. But

$$D[f[-x], x] = \frac{1}{x} + 0$$

because the derivative of the constant $i\pi$ is 0.

This tells you that $-f'[-x] = 1/x$, and this is the same as $f'[-x] = -1/x = 1/(-x)$.

This explains why $f'[x] = 1/x$ even when x is negative.

Tutorials

■ T.1) Formulas for the solutions of certain differential equations by separating and integrating

You might remember a differential equation of the following type from the first lesson on differential equations early in the course. All the problems from that lesson can be handled by separating and integrating.

T.1.a)

> For given positive constants a, b, and c, separate and integrate to come up with a formula for the solution of
>
> $$y'[x] = c - a\, y[x] \qquad \text{given that } y[0] = b.$$
>
> Confirm your formula with DSolve.
>
> What is the limiting value
>
> $$\lim_{x \to \infty} y[x]?$$

Answer: Rewrite $y'[x] = c - a\,y[x]$ as

$$\frac{y'[x]}{c - a\,y[x]} = 1$$

and integrate:

In[1]:=
```
Clear[y,t,a,b,c,x]
left = Integrate[y'[t]/(c - a y[t]),{t,0,x}]/.y[0]->b
```

Out[1]=
$$\frac{\text{Log[a b - c]}}{a} - \frac{\text{Log[-c + a y[x]]}}{a}$$

If $c > a\,b$, then this involes the log of a negative number, but logs of negative numbers are no bother to you.

Fix it up:

In[2]:=
```
Clear[p,q]
simpleleft = left/.(Log[p_]/a - Log[q_]/a)->(Log[p/q]/a)
```

Out[2]=

$$\frac{\text{Log}\left[\dfrac{a\ b\ -\ c}{-c\ +\ a\ y[x]}\right]}{a}$$

In[3]:=
```
right = Integrate[1,{t,0,x}]
```

Out[3]=

x

To start to solve for $y[x]$, multiply both sides by a:

In[4]:=
```
newleft = a simpleleft
```

Out[4]=

$$\text{Log}\left[\frac{a\ b\ -\ c}{-c\ +\ a\ y[x]}\right]$$

In[5]:=
```
newright = a right
```

Out[5]=

a x

Exponentiate both sides:

In[6]:=
```
betterleft = E^newleft
```

Out[6]=

$$\frac{a\ b\ -\ c}{-c\ +\ a\ y[x]}$$

In[7]:=
```
betterright = E^newright
```

Out[7]=

$$E^{a\ x}$$

Here comes the formula:

In[8]:=
```
ysolved = Solve[betterleft == betterright,y[x]];
y[x_] = ysolved[[1,1,2]]
```

Out[8]=

$$\frac{a\ b\ -\ c\ +\ c\ E^{a\ x}}{a\ E^{a\ x}}$$

Now confirm the formula with DSolve:

In[9]:=
```
Clear[y,x,c,a,b,Derivative]
DSolve[{y'[x] == c - a y[x],y[0] == b},y[x],x]
```

Out[9]=

$$\{\{y[x] \; \rightarrow \; \frac{a \; b \; - \; c \; + \; c \; E^{a \; x}}{a \; E^{a \; x}}\}\}$$

Good.

To investigate the limiting behavior of $y[x]$, look at the quotient of the dominant terms

$$\frac{c \, e^{ax}}{a \, e^{ax}} = \frac{c}{a}.$$

This tells you that

$$\lim_{x \to \infty} y[x] = \frac{c}{a}.$$

T.1.b)

A spherical iceball whose initial ($t = 0$) radius is 6 inches melts so that its radius decreases at a rate proportional to the area of its surface.

Given that it takes half an hour for the radius to decrease from 6 inches to 3 inches, come up with a formula that measures the radius of the iceball at any time $t > 0$.

Confirm your formula with DSolve.

Use your formula to determine how long it will take for the radius to decrease to 0.1 inch.

Answer: All t measurements are in hours. The surface area of a ball of radius r is proportional to r^2. The upshot: If radius$[t]$ is the radius of the iceball at time t, then saying the radius decreases at a rate proportional to the area of its surface is the same as saying radius$'[t]$ is proportional to radius$[t]^2$. This is neat because it gets around the problem of calculating surface area of a sphere.

This means

$$\text{radius}'[t] = K \, \text{radius}[t]^2$$

for some (yet to be determined) constant of proportionality K. Separate the variables t and radius$[t]$ by rewriting as

$$\frac{\text{radius}'[t]}{\text{radius}[t]^2} = K,$$

and integrate from 0 to t, all the while remembering that radius$[0] = 6$:

In[10]:=
```
Clear[K,r,t,x,radius]
left = Integrate[radius'[x]/radius[x]^2,{x,0,t}]/.radius[0]->6
```

Out[10]=

$$\frac{1}{6} - \frac{1}{\text{radius}[t]}$$

In[11]:=
```
right = Integrate[K,{x,0,t}]
```
Out[11]=
```
K t
```

Here comes the first shot at a formula for radius[*t*]:

In[12]:=
```
radsolved = Solve[left == right,radius[t]]
```
Out[12]=

$$\{\{\text{radius}[t] \rightarrow \frac{6}{1 - 6\,K\,t}\}\}$$

In[13]:=
```
Clear[prelimformula]
prelimformula[t_] = radius[t]/.radsolved[[1]]
```
Out[13]=

$$\frac{6}{1 - 6\,K\,t}$$

Confirm with DSolve:

In[14]:=
```
Clear[radius,t,K,Derivative]
DSolve[{radius'[t] == K radius[t]^2,radius[0] == 6},radius[t],t]
```
Out[14]=

$$\{\{\text{radius}[t] \rightarrow \frac{1}{\frac{1}{6} - K\,t}\}\}$$

To see that the formulas are the same, take the output from DSolve and multiply the top and bottom by 6.

Now use the information that radius[1/2] = 3.

In[15]:=
```
Solve[3 == prelimformula[1/2],K]
```
Out[15]=

$$\{\{K \rightarrow -(\frac{1}{3})\}\}$$

Now you've got your hands on the formula:

In[16]:=
```
radius[t_] = prelimformula[t]/.K->-1/3
```
Out[16]=

$$\frac{6}{1 + 2\,t}$$

To find how long it takes for the radius to shrink to 0.1 inches, solve:

In[17]:=
```
Solve[0.1 == radius[t]]
```

Out[17]=
    {{t -> 29.5}}

After a little more than one day, the iceball is nearly gone.

T.1.c) Here is the differential equation of the harmonic oscillator:
$$y''[x] + K\,y[x] = 0 \qquad \text{with } y[0] = a \text{ and } y'[0] = b.$$

In principle you can separate the variables by rewriting the differential equation as
$$\frac{y''[x]}{y[x]} = -K$$

and integrating.

> Try it and report your findings.

Answer: Here you go:

In[18]:=
    Clear[y,t,a,b,c,x]
    left = Integrate[y''[t]/y[t],{t,0,x}]/.{y[0]->a,y'[0]->b}

Out[18]=

$$-\left(\frac{b}{a}\right) + \text{Integrate}\left[\frac{y'[t]^2}{y[t]^2}, \{t, 0, x\}\right] + \frac{y'[x]}{y[x]}$$

Abort before your machine crashes.

This is not successful because neither *Mathematica* nor anything else can handle $\int_0^x y''[t]/y[t]\,dt$. Generally, separating and integrating works only with special differential equations involving only $y[x]$ and $y'[x]$. Separating and integrating is not worth trying when higher derivatives are involved.

■ T.2) Using integration by parts to do integration by iteration

T.2.a)
> Use the integration by parts formula and iteration to prepare a little table of the values of
> $$\int_0^1 x^n\,e^x\,dx \qquad \text{for } n = 0, 1, 2, 3, \ldots, 15$$
> by evaluating only one actual integral.

Answer: The integration by parts formula is
$$\int_0^1 u[x]\,v'[x]\,dx = u[x]\,v[x]\Big|_0^1 - \int_0^1 v[x]\,u'[x]\,dx.$$

Make the assignments: $u[x] = x^n$ and $v'[x] = e^x$. This gives $u'[x] = n\,x^{n-1}$ and $v[x] = e^x$. With these assignments,

$$\int_0^1 x^n\,e^x\,dx = \int_0^1 u[x]\,v'[x]\,dx$$

$$= u[x]\,v[x]\,\big[_0^1 - \int_0^1 v[x]\,u'[x]\,dx$$

$$= x^n\,e^x\,\big[_0^1 - \int_0^1 e^x\,n\,x^{n-1}\,dx$$

$$= e - n\int_0^1 x^{n-1}\,e^x\,dx.$$

Accordingly,

$$\int_0^1 x^n\,e^x\,dx = e - n\int_0^1 x^{n-1}\,e^x\,dx.$$

Something great just happened. Letting

$$\text{Int}[n] = \int_0^1 x^n\,e^x\,dx,$$

you see that the formula immediately above takes the form

$$\text{Int}[n] = e - n\,\text{Int}[n-1].$$

This iteration formula will let you build the whole table on the basis of evaluating just one lonesome integral:

In[19]:=
```
Clear[n,x,Int]
```

Integrate to find $\text{Int}[0] = \int_0^1 x^0\,e^x\,dx$:

In[20]:=
```
Int[0] = Integrate[E^x,{x,0,1}]
```

Out[20]=
```
-1 + E
```

Enter the iteration formula:

In[21]:=
```
Int[n_] := Int[n] = E - n Int[n - 1]
```

Now you can get the whole table

$$\{\text{Int}[0], \text{Int}[1], \text{Int}[2], \ldots, \text{Int}[15]\}$$

$$= \left\{\int_0^1 e^x\,dx, \int_0^1 x\,e^x\,dx, \int_0^1 x^2\,e^x\,dx, \ldots, \int_0^1 x^{15}\,e^x\,dx\right\}$$

with no additional work:

In[22]:=
```
ColumnForm[Table[Expand[{"Int"[n],Int[n]}],{n,0,15}]]
```

Out[22]=
```
{Int[0],  -1 + E}
{Int[1],  1}
{Int[2],  -2 + E}
{Int[3],  6 - 2 E}
{Int[4],  -24 + 9 E}
{Int[5],  120 - 44 E}
{Int[6],  -720 + 265 E}
{Int[7],  5040 - 1854 E}
{Int[8],  -40320 + 14833 E}
{Int[9],  362880 - 133496 E}
{Int[10],  -3628800 + 1334961 E}
{Int[11],  39916800 - 14684570 E}
{Int[12],  -479001600 + 176214841 E}
{Int[13],  6227020800 - 2290792932 E}
{Int[14],  -87178291200 + 32071101049 E}
{Int[15],  1307674368000 - 481066515734 E}
```

Now you can read off, for instance, $\int_0^1 x^9 e^x \, dx = \text{Int}[9] = 362880 - 133496 \, e$. You can try for machine accuracy:

In[23]:=
```
N[Int[9]]
```

Out[23]=
```
0.249028
```

A lot more accuracy:

In[24]:=
```
N[Int[9],30]
```

Out[24]=
```
0.249028031297260343306372
```

T.2.b)

> Use the integration by parts formula to help prepare a table of the values of $\int_0^{\pi/2} \sin[x]^n \, dx$ for $n = 1, 2, 3, \ldots, 18$ by evaluating only two actual integrals. Comment on the calculational advantage of iteration over direct calculation in this situation.

Answer: Here's a little trick:

$$\int_0^{\pi/2} \sin[x]^n \, dx = \int_0^{\pi/2} \sin[x]^{n-1} \sin[x] \, dx.$$

The integration by parts formula is

$$\int_0^{\pi/2} u[x] \, v'[x] \, dx = u[x] \, v[x] \Big[_0^{\pi/2} - \int_0^{\pi/2} v[x] \, u'[x] \, dx.$$

Make the assignments: $u[x] = \sin[x]^{n-1}$ and $v'[x] = \sin[x]$. This gives

$$u'[x] = (n - 1) \sin[x]^{n-2} \cos[x]$$

and

$$v[x] = -\cos[x].$$

With these assignments,

$$\int_0^{\pi/2} \sin[x]^n \, dx = \int_0^{\pi/2} \sin[x]^{n-1} \sin[x] \, dx$$

$$= \int_0^{\pi/2} u[x] \, v'[x] \, dx$$

$$= u[x] \, v[x] \Big[_0^{\pi/2} - \int_0^{\pi/2} v[x] \, u'[x] \, dx$$

$$= \sin[x]^{n-1} \, (-\cos[x]) \, \Big[_0^{\pi/2} - \int_0^{\pi/2} (-\cos[x])$$
$$\times \, (n-1) \sin[x]^{n-2} \cos[x] \, dx$$

$$= 0 + (n-1) \int_0^{\pi/2} \sin[x]^{n-2} \cos[x]^2 \, dx$$

$$= (n-1) \int_0^{\pi/2} \sin[x]^{n-2} \left(1 - \sin[x]^2\right) \, dx$$

$$= (n-1) \int_0^{\pi/2} \sin[x]^{n-2} \, dx - (n-1) \int_0^{\pi/2} \sin[x]^n \, dx.$$

The upshot:

$$\int_0^{\pi/2} \sin[x]^n \, dx = (n-1) \int_0^{\pi/2} \sin[x]^{n-2} \, dx - (n-1) \int_0^{\pi/2} \sin[x]^n \, dx.$$

This looks bad, but it feels good because when you put

$$\text{Int}[n] = \int_0^{\pi/2} \sin[x]^n \, dx,$$

then:

In[25]:=
```
Clear[Int,n]
equation = Int[n] == (n - 1) Int[n - 2] - (n - 1) Int[n]
```

Out[25]=
```
Int[n] == (-1 + n) Int[-2 + n] - (-1 + n) Int[n]
```

This gives:

In[26]:=
```
ExpandAll[Solve[equation,Int[n]]]
```

Out[26]=

$$\{\{\text{Int}[n] \; \text{->} \; \text{Int}[-2 + n] \; - \; \frac{\text{Int}[-2 + n]}{n}\}\}$$

So:

In[27]:=
```
Int[n_] := Int[n] = Int[n - 2] - Int[n - 2]/n
```

Calculate Int[1] and Int[2]:

In[28]:=
```
Int[1] = Integrate[Sin[x]^1,{x,0,Pi/2}]
```

Out[28]=
```
1
```

In[29]:=
```
Int[2] = Integrate[Sin[x]^2,{x,0,Pi/2}]
```

Out[29]=
$$\frac{Pi}{4}$$

Now you can get the whole table
$$\{Int[1], Int[2], \ldots, Int[18]\}$$
$$= \left\{ \int_0^{\pi/2} \sin[x] \, dx, \ \int_0^{\pi/2} \sin[x]^2 \, dx, \ldots, \int_0^{\pi/2} \sin[x]^{18} \, dx \right\} :$$

In[30]:=
```
ColumnForm[Table[Expand[{"Int"[n],Int[n]}],{n,1,18}]]
```

Out[30]=

ColumnFormForm[{{Int[1], 1}, {Int[2], $\frac{Pi}{4}$}, {Int[3], $\frac{2}{3}$},

{Int[4], $\frac{3\ Pi}{16}$}, {Int[5], $\frac{8}{15}$}, {Int[6], $\frac{5\ Pi}{32}$}, {Int[7], $\frac{16}{35}$},

{Int[8], $\frac{35\ Pi}{256}$}, {Int[9], $\frac{128}{315}$}, {Int[10], $\frac{63\ Pi}{512}$},

{Int[11], $\frac{256}{693}$}, {Int[12], $\frac{231\ Pi}{2048}$}, {Int[13], $\frac{1024}{3003}$},

{Int[14], $\frac{429\ Pi}{4096}$}, {Int[15], $\frac{2048}{6435}$}, {Int[16], $\frac{6435\ Pi}{65536}$},

{Int[17], $\frac{32768}{109395}$}, {Int[18], $\frac{12155\ Pi}{131072}$}}]

That was quick. Spot check:

In[31]:=
```
Integrate[Sin[x]^5,{x,0,Pi/2}] == Int[5]
```

Out[31]=
```
True
```

Good. The whole table was calculated in slightly over two seconds on a Macintosh IIcx, but the evaluation of the individual spot check Int[5] $= \int_0^{\pi/2} \sin[x]^5 \, dx$ took more than six seconds on the same machine. This shows off the calculational advantage of the iteration.

■ **T.3)** Using the complex exponential to help understand the *Mathematica* output from the Solve instruction

This problem appears only in the electronic version.

Give It a Try

Experience with the starred (⋆) problems will be especially beneficial for understanding later lessons.

■ **G.1)** **Separating and integrating**

G.1.a)

> Use separation of variables to get a formula for the solution of the differential equation
>
> $$y'[x] = \sin[x]\, y[x]^2 \qquad \text{with } y[0] = 4.$$
>
> Use DSolve to confirm your formula.
>
> Plot your solution and $\sin[x]$ on the same axes, discuss the relations between the two plots, and try to account for the relationship you spot.

G.1.b) A certain living cell of constant density does not change shape or composition as it grows. Consequently, the area of its surface is proportional to the square of any of its linear dimensions and its weight is proportional to the cube of any of its linear dimensions. This cell takes in nutrients through its surface; so if $W[t]$ is its weight at time t, then $W'[t]$ is proportional to the area of its surface.

> Explain why $W'[t] = K\, W[t]^{2/3}$ for some proportionality constant K. Use separation of variables to get a formula for the solution of this assuming $W[0] = 1$, and interpret the solution.

G.1.c) A bathtub is filled with water to a depth of 9 inches. At time $t = 0$, the plug is pulled out and the water begins to run out of the tub. The depth, $h[t]$ measured in inches, is decreasing at a rate given by

$$\frac{dh[t]}{dt} = -\left(\frac{1}{\pi}\right)\sqrt{h[t]} \qquad \text{(with } t \text{ in seconds).}$$

> Use separation of variables to get a formula for the solution of this differential equation. Confirm your formula with DSolve.
>
> How long does it take for all but a drop or two of the water to run out of the tub?

■ G.2) Integration by parts

G.2.a) Calculate the following integrals exactly (no approximations) by the method of integration by parts; check with *Mathematica*.

G.2.a.i)
$$\int_1^e \log[x]\,dx$$

G.2.a.ii)
$$\int_0^t \text{SinIntegral}[x]\,dx$$

G.2.a.iii)
$$\int_0^t \text{FresnelS}\left[\sqrt{\frac{2}{\pi}}\,x\right]\,dx$$

G.2.a.iv)
$$\int_0^{1/2} x\,\sin[\pi\,x]\,dx$$

G.2.a.v)
$$\int_0^{\infty} x\,e^{-3x}\,dx$$

■ G.3) Chemical reaction model and the spread of infection model

G.3.a) Two chemicals A and B react with each other and, in the process, one molecule from A bonds with one molecule from B to form a new compound. Let

a be the number of molecules of A present at the start

and let

b be the number of molecules of B present at the start.

Let $y[t]$ be the number of molecules of the new compound present t seconds after the reaction starts.

At any time t, there are $(a - y[t])$ molecules of A available for reaction and there are $(b - y[t])$ molecules of B available for reaction.

Because the reaction requires collision of an A-molecule and a B-molecule, it makes sense to assume that $y'[t]$ is proportional to both $(a - y[t])$ and $(b - y[t])$ in the sense that

$$y'[t] = k\,(a - y[t])\,(b - y[t])$$

for some proportionality constant k.

This differential equation says that the rate of reaction is proportional to the product of the concentrations of the reactants. Fancy folks call this the Law of Mass Action.

G.3.a.i) What is the value of $y[0]$ at the instant the reaction starts?

G.3.a.ii) Use the technique of separating and integrating to find a formula for $y[t]$ in terms of a, b, and k. Confirm your formula with DSolve.

What is the limiting behavior of $y[t]$ if $a > b$?

What if $b > a$?

For more on chemical reaction models, see the book by J. L. Latham, *Elementary Reaction Kinetics*, Butterworths, London, 1964. Pay special attention to the material near page 100.

G.3.b) You have a closed population of P equally suspectible infection-free individuals. One additional infected individual is introduced. Put

$$y[t] = \text{ the number of infected individuals at time } t$$

and

$$x[t] = \text{ the others.}$$

This gives you

$$x[t] + y[t] = P + 1.$$

G.3.b.i) Why is it reasonable to assume that the rate $y'[t]$ at which the infected population is growing is proportional to both $y[t]$ and to $x[t]$?

Why does this assumption result in the relationship

$$y'[t] = k\,y[t]\,(P + 1 - y[t])?$$

G.3.b.ii) Given $y[0] = 1$, use the technique of separating and integrating to come up with a formula for $y[t]$ in terms of t, k, and P. Confirm your formula with DSolve.

G.3.b.iii) Find the value of k under the assumption that $P = 1000$ and that at time $t = 5$, you know that 200 individuals are infected. Then plot $y[t]$. How large is $y[t]$ when the infection is spreading most rapidly?

■ G.4) Tables of integrals via iteration

G.4.a.i) Use the integration by parts formula to calculate

$$\int_1^e \log[x]\,dx$$

by setting

$$u[x] = \log[x] \qquad \text{and} \qquad v'[x] = 1.$$

G.4.a.ii) Calculate an expression for

$$\int_1^e \log[x]^k\,dx$$

in terms of

$$\int_1^e \log[x]^{k-1}\,dx$$

Use this expression and the result from part G.4.a.i) above to build a table of the values of

$$\int_1^e \log[x]^k\,dx \qquad \text{for } k = 1, 2, 3, \ldots, 15$$

evaluating only one actual integral.

G.4.b) Use integration by parts to prepare a table of the values of

$$\int_0^\pi \cos[x]^n\,dx \qquad \text{for } n = 1, 2, 3, \ldots, 20$$

by evaluating only two actual integrals.

■ G.5) Meet sinh[x] and cosh[x]*

G.5.a) Often called the hyperbolic sine, the formula for $\sinh[x]$ is:

$$\sinh[x] = \frac{e^x - e^{-x}}{2}.$$

The groovy folks call sinh[x] by the name "cinch x."

Here's a plot of sinh[x] along with its part-time companions $e^x/2$ and $-e^{-x}/2$:

In[1]:=
```
Clear[x]
Plot[{Sinh[x],
(E^x)/2,-(E^(-x))/2},{x,-2.5,2.5},
PlotStyle->{{Blue,Thickness[0.01]},Red,Red},
PlotRange->All,AxesLabel->{"x",""}];
```

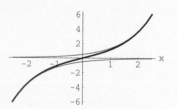

sinh[x] switches its allegiance from $-e^{-x}/2$ to $e^x/2$.

Use the formula for sinh[x],

$$\sinh[x] = \frac{e^x - e^{-x}}{2}$$

and what you know about exponential growth to account for why the plot turned out the way it did.

G.5.b) Often called the hyperbolic cosine, cosh[x] is given by the formula:

$$\cosh[x] = \frac{e^x + e^{-x}}{2}.$$

The groovy folks call cosh[x] by the name "cosh x."

Here's a plot of cosh[x] along with its part-time companions $e^x/2$ and $e^{-x}/2$:

In[2]:=
```
Clear[x]
Plot[{Cosh[x],
(E^x)/2,+(E^(-x))/2},{x,-2.5,2.5},
PlotStyle->{{Blue,Thickness[0.01]},Red,Red},
PlotRange->All,AxesLabel->{"x",""}];
```

Some folks call the plot of cosh[x] by the name "catenary." An electric wire strung between two telephone poles hangs in the shape of a catenary.

cosh[x] switches its allegiance from $e^{-x}/2$ to $e^x/2$.

Use the formula for cosh[x],

$$\cosh[x] = \frac{e^x + e^{-x}}{2}$$

and what you know about exponential growth to account for why the plot turned out the way it did.

G.5.c) Look at:

In[3]:=
```
Clear[x]; D[Sinh[x],x]
```

Out[3]=
```
Cosh[x]
```

> Use the formulas
> $$\sinh[x] = \frac{e^x - e^{-x}}{2} \qquad \text{and} \qquad \cosh[x] = \frac{e^x + e^{-x}}{2}$$
> to explain why the derivative of $\sinh[x]$ is $\cosh[x]$.

G.5.d) Look at:

In[4]:=
```
Clear[x]; D[Cosh[x],x]
```

Out[4]=
```
Sinh[x]
```

> Use the formulas
> $$\sinh[x] = \frac{e^x - e^{-x}}{2} \qquad \text{and} \qquad \cosh[x] = \frac{e^x + e^{-x}}{2}$$
> to explain why the derivative of $\cosh[x]$ is $\sinh[x]$.

G.5.e) Recall from above:

In[5]:=
```
Clear[x]; D[Sin[x],x]
```

Out[5]=
```
Cos[x]
```

In[6]:=
```
Clear[x]; D[Sinh[x],x]
```

Out[6]=
```
Cosh[x]
```

In[7]:=
```
Clear[x]; D[Cos[x],x]
```

Out[7]=
```
-Sin[x]
```

In[8]:=
```
Clear[x]; D[Cosh[x],x]
```

Out[8]=
```
Sinh[x]
```

There is something sneaky going on here. There has got to be a connection between the trig functions sin[x], cos[x] and the hyperbolic functions sinh[x], cosh[x]. Here is one of the connections:

In[9]:=
```
Expand[Sinh[I x],Trig->True]
```

Out[9]=
```
I Sin[x]
```

Cowabunga! Apparently

$$\sinh[i\,x] = i\,\sin[x].$$

Here is the other link:

In[10]:=
```
Expand[Cosh[I x],Trig->True]
```

Out[10]=
```
Cos[x]
```

Cowagoopa! Apparently

$$\cosh[i\,x] = \cos[x].$$

These are amazing facts because they give a concrete link among sin[x], sinh[x], cos[x], and cosh[x]. And this link is not too hard to understand. Just remember that:

$$e^{ix} = \cos[x] + i\,\sin[x]$$
$$e^{-ix} = \cos[-x] + i\,\sin[-x] = \cos[x] - i\,\sin[x].$$

Solve for sin[x] and cos[x] in terms of e^{ix} and e^{-ix}:

In[11]:=
```
Clear[x]
Solve[{E^(I x) == Cos[x] + I Sin[x],
E^(-I x) == Cos[x] - I Sin[x]}, {Sin[x],Cos[x]}]
```

Out[11]=
```
                                         -I x    I x
           I -I x   I  I x              E       E
{{Sin[x] -> - E     - - E    , Cos[x] -> ----- + -----}}
           2         2                     2       2
```

This tells you

$$\sin[x] = i\,\frac{e^{-ix} - e^{ix}}{2}$$

and

$$\cos[x] = \frac{e^{-ix} + e^{ix}}{2}.$$

These help to explain:

In[12]:=
```
Expand[Sinh[I x],Trig->True]
```

Out[12]=
  I Sin[x]

To explain why $\sinh[i\,x] = i\,\sin[x]$, just use

$$\sin[x] = i\,\frac{e^{-ix} - e^{ix}}{2}$$

and multiply both sides by i to get:

$$
\begin{aligned}
i\,\sin[x] &= \frac{i^2\left(e^{-ix} - e^{ix}\right)}{2}\\[4pt]
&= \frac{(-1)\left(e^{-ix} - e^{ix}\right)}{2}\\[4pt]
&= \frac{\left(-e^{-ix} + e^{ix}\right)}{2}\\[4pt]
&= \frac{\left(e^{ix} - e^{-ix}\right)}{2}\\[4pt]
&= \sinh[i\,x]
\end{aligned}
$$

because $\sinh[x] = \left(e^x - e^{-x}\right)/2$.

> Now you take over and explain why it turns out that $\cosh[i\,x] = \cos[x]$:

■ G.6) The gamma function

This problem appears only in the electronic version.

■ G.7) "The algebra of [the complex] exponentials is much easier than that of sines and cosines," said Richard Feynman★

Richard Feynman, a long time professor at Cal Tech who won the Nobel Prize in 1965, was one of the world's great theoretical physicists. His whole life, scientific and otherwise, was one big search for adventure. Bongo drums and topless bars were two of his favorite extracurricular activities. Shortly before his death (from cancer in 1988, two weeks after he taught his last class at Cal Tech), Feynman jumped into the focus of the national media when he dropped a ring of rubber into a glass of cold water and pulled it out, revealing that the experiment had misshapened the rubber ring. In so doing, he demonstrated to the nation the cause of the explosion of the space shuttle Challenger in 1986.

The great American physicist Richard Feynman once said, "The algebra of [the complex] exponentials is much easier than that of sines and cosines."

Here is what Feynman meant:

Darn few folks go around knowing what $\sin[3\,x]$ and $\cos[3\,x]$ are in terms of $\sin[x]$ and $\cos[x]$. But all folks like you and Feynman know that

$$\cos[3\,x] + i \, \sin[3\,x] = e^{i3x} = \left(e^{ix}\right)^3 = (\cos[x] + i \, \sin[x])^3$$

and this is just:

In[13]:=
```
Expand[(Cos[x] + I Sin[x])^3]
```

Out[13]=

$\text{Cos[x]}^3 + 3 \text{ I Cos[x]}^2 \text{ Sin[x]} - 3 \text{ Cos[x] Sin[x]}^2 - \text{ I Sin[x]}^3$

This tells you

$$\cos[3\,x] = \cos[x]^3 - 3\,\cos[x]\,\sin[x]^2$$

and

$$\sin[3\,x] = 3\,\cos[x]^2\,\sin[x] - \sin[x]^3.$$

You can automate this way of generating multiple angle identities through the formulas

$$\cos[m\,x] + i\,\sin[m\,x] = (\cos[x] + i\,\sin[x])^m$$

and

$$\cos[m\,x] - i\,\sin[m\,x] = (\cos[x] - i\,\sin[x])^m\,.$$

They both come from the basic formula $\cos[x] + i\,\sin[x] = e^{ix}$.

These formulas were first observed by Abraham DeMoivre, a gifted French mathematician banished to England as a result of the French revolution who supported himself by supplying information on games of chance to rich gentleman gamblers.

To use these formulas to generate identities for $\sin[m\,x]$, take the formulas

$$\cos[m\,x] + i\,\sin[m\,x] = (\cos[x] + i\,\sin[x])^m$$

and

$$\cos[m\,x] - i\,\sin[m\,x] = (\cos[x] - i\,\sin[x])^m$$

and subtract the second from the first to get

$$2\,i\,\sin[m\,x] = (\cos[x] + i\,\sin[x])^m - (\cos[x] - i\,\sin[x])^m\,.$$

Multiply both sides by $(-i/2)$ to get

$$\sin[m\,x] = \left(\frac{-i}{2}\right)\,2\,i\,\sin[m\,x] = \left(\frac{-i}{2}\right)\left((\cos[x] + i\,\sin[x])^m - (\cos[x] - i\,\sin[x])^m\right).$$

Now turn *Mathematica* loose: $\sin[2\,x]$ is:

In[14]:=
```
m = 2;
Expand[(-I/2)((Cos[x] + I Sin[x])^m - (Cos[x] - I Sin[x])^m)]
```

Out[14]=
```
 2 Cos[x] Sin[x]
```

Looks good and feels good. $\sin[3\,x]$ is:

In[15]:=
```
 m = 3;
 Expand[(-I/2)((Cos[x] + I Sin[x])^m - (Cos[x] - I Sin[x])^m)]
```

Out[15]=
```
            2            3
 3 Cos[x]  Sin[x] - Sin[x]
```

Good. $\sin[4\,x]$ is:

In[16]:=
```
 m = 4;
 Expand[(-I/2)((Cos[x] + I Sin[x])^m - (Cos[x] - I Sin[x])^m)]
```

Out[16]=
```
          3                      3
 4 Cos[x]  Sin[x] - 4 Cos[x] Sin[x]
```

Good. $\sin[20\,x]$ is:

In[17]:=
```
 m = 20;
 Expand[(-I/2)((Cos[x] + I Sin[x])^m - (Cos[x] - I Sin[x])^m)]
```

Out[17]=
```
           19                 17       3              15       5
 20 Cos[x]   Sin[x] - 1140 Cos[x]  Sin[x]  + 15504 Cos[x]  Sin[x]  -

              13       7               11       9               9       11
 77520 Cos[x]   Sin[x]  + 167960 Cos[x]  Sin[x]  - 167960 Cos[x]  Sin[x]   +

             7       13              5       15             3       17
 77520 Cos[x]  Sin[x]   - 15504 Cos[x]  Sin[x]   + 1140 Cos[x]  Sin[x]   -

              19
 20 Cos[x] Sin[x]
```

Calculus&*Mathematica* students are among the few mortals who have seen this last formula.

Good. Play with some other formulas for $\sin[m\,x]$ for positive integers m if you like.

It's a good thing your trig teacher didn't have a copy of *Mathematica*, and it's too bad you didn't have one. You and *Mathematica* could have really shaken up a traditionally taught trig course.

G.7.a)

Take the formulas
$$\cos[m\,x] + i\,\sin[m\,x] = (\cos[x] + i\,\sin[x])^m$$
and
$$\cos[m\,x] - i\,\sin[m\,x] = (\cos[x] - i\,\sin[x])^m$$

and add them to derive a formula $\cos[m\,x]$. Then turn your formula over to *Mathematica* to give formulas for $\cos[2\,x]$, $\cos[3\,x]$, $\cos[9\,x]$, and $\cos[20\,x]$ in terms of powers of $\sin[x]$ and $\cos[x]$.

■ G.8) Error propagation via iteration: Against you and for you

Calculus&*Mathematica* thanks Professor Julian Palmore of the University of Illinois for suggesting this problem.

You can use integration by parts to prepare a table of the values of $\int_0^1 x^n\,e^x\,dx$ for various values of n's. Start out by setting

$$\text{Int}[n] = \int_0^1 x^n\,e^x\,dx$$

and then integrate by parts to learn

$$\text{Int}[n] = e - n\,\text{Int}[n-1].$$

Use this iteration by entering the exact value of

$$\text{Int}[0] = \int_0^1 x^0\,e^x\,dx = \int_0^1 e^x\,dx = e - 1.$$

In[18]:=
```
Clear[x,Int,n]; Int[0] = E - 1
```
Out[18]=
```
-1 + E
```

Type the iteration formula:

In[19]:=
```
Int[n_] := Int[n] = Expand[E - n Int[n - 1]]
```

Make a table of exact values:

In[20]:=
```
ColumnForm[Table[{"Int"[n],Int[n]},{n,0,12}]]
```

Out[20]=
```
{Int[0], -1 + E}
{Int[1], 1}
{Int[2], -2 + E}
{Int[3], 6 - 2 E}
{Int[4], -24 + 9 E}
{Int[5], 120 - 44 E}
{Int[6], -720 + 265 E}
{Int[7], 5040 - 1854 E}
{Int[8], -40320 + 14833 E}
{Int[9], 362880 - 133496 E}
{Int[10], -3628800 + 1334961 E}
{Int[11], 39916800 - 14684570 E}
{Int[12], -479001600 + 176214841 E}
```

Now see what happens when you do the same thing, but instead of entering the exact value $\text{Int}[0] = e - 1$, you enter a rounded off decimal approximation of $e - 1$:

In[21]:=
```
N[E - 1]
```

Out[21]=
```
1.71828
```

In[22]:=
```
Clear[x,IInt,n]; IInt[0] = 1.71828
```

Out[22]=
```
1.71828
```

Type the iteration formula:

In[23]:=
```
IInt[n_] := IInt[n] = Expand[E - n IInt[n - 1]]
```

Make a table of values:

In[24]:=
```
ColumnForm[Table[{"Int"[n],N[IInt[n],20]},{n,0,12}]]
```

Out[24]=
```
{Int[0],  1.71828}
{Int[1],  1.000001828459045}
{Int[2],  0.718278171540955}
{Int[3],  0.5634473138361802}
{Int[4],  0.4644925731143204}
{Int[5],  0.3958189628874322}
{Int[6],  0.3433680511343482}
{Int[7],  0.3147054705186747}
{Int[8],  0.200638064314262}
{Int[9],  0.912539249635302}
{Int[10], -6.407110667787493}
{Int[11], 73.19649916887284}
{Int[12], -875.639708280563}
```

Make a table comparing the exact value and the values of IInt[n]:

In[25]:=
```
ColumnForm[Table[{"Int"[n],N[Int[n],20],N[IInt[n],20]},{n,0,12}]]
```

Out[25]=
```
{Int[0],  1.7182818284590452354,  1.71828}
{Int[1],  1.,  1.000001828459045}
{Int[2],  0.7182818284590452354,  0.718278171540955}
{Int[3],  0.563436343081909529,  0.5634473138361802}
{Int[4],  0.464536456131407118,  0.4644925731143204}
{Int[5],  0.39559954780200964,  0.3958189628874322}
{Int[6],  0.3446845416469874,  0.3433680511343482}
{Int[7],  0.305490036930134,  0.3147054705186747}
{Int[8],  0.274361533017976,  0.200638064314262}
{Int[9],  0.24902803129726,  0.912539249635302}
{Int[10], 0.2280015154864,  -6.407110667787493}
{Int[11], 0.210265158108,  73.19649916887284}
{Int[12], 0.19509993116,  -875.639708280563}
```

For small n's, the exact values Int[n] are very close to the approximate values Int[n], but for larger n's, there are dramatic discrepancies. Here's what's happening: Suppose exact0 stands for the exact value of Int[0]:

In[26]:=
```
Clear[x,Int,n]; Int[0] = exact0
```

Out[26]=
```
exact0
```

In terms of exact0, the exact values of Int[0], Int[1], . . ., Int[15] are:

In[27]:=
```
Int[n_] := Int[n] = Expand[E - n Int[n - 1]]
```

In[28]:=
```
ColumnForm[Table[{"Int"[n],N[Int[n],20]},{n,0,12}]]
```

Out[28]=
```
{Int[0], exact0}
{Int[1], 2.7182818284590452354 - 1. exact0}
{Int[2], -2.7182818284590452354 + 2. exact0}
{Int[3], 10.873127313836180941 - 6. exact0}
{Int[4], -40.774227426885567853 + 24. exact0}
{Int[5], 206.58941896288743789 - 120. exact0}
{Int[6], -1236.8182319488655821 + 720. exact0}
{Int[7], 8660.44590547051812 - 5040. exact0}
{Int[8], -69280.84896193568591 + 40320. exact0}
{Int[9], 623530.3589392496323 - 362880. exact0}
```

$\{\text{Int}[10], -6.235300871110667864 \ 10^6 + 3.6288 \ 10^6 \ \text{exact0}\}$

$\{\text{Int}[11], 6.858831230049917496 \ 10^7 - 3.99168 \ 10^7 \ \text{exact0}\}$

$\{\text{Int}[12], -8.23059744887708271 \ 10^8 + 4.790016 \ 10^8 \ \text{exact0}\}$

Now watch what happens when you build in an error = error0 into an approximate value (exact0 + error0) for Int[0]:

In[29]:=
```
Clear[x,Int,n]; Int[0] = exact0 + error0
```

Out[29]=
```
error0 + exact0
```

In[30]:=
```
Int[n_] := Int[n] = Expand[E - n Int[n - 1]]
```

The corresponding values of Int[0], Int[1], . . ., and Int[12] are:

In[31]:=
```
ColumnForm[Table[{"Int"[n],N[Int[n],12]},{n,0,12}]]
```

Out[31]=
```
{Int[0], error0 + exact0}
{Int[1], 2.71828182846 - 1. error0 - 1. exact0}
{Int[2], -2.71828182846 + 2. error0 + 2. exact0}
{Int[3], 10.8731273138 - 6. error0 - 6. exact0}
{Int[4], -40.7742274269 + 24. error0 + 24. exact0}
{Int[5], 206.589418963 - 120. error0 - 120. exact0}
{Int[6], -1236.81823195 + 720. error0 + 720. exact0}
{Int[7], 8660.44590547 - 5040. error0 - 5040. exact0}
{Int[8], -69280.8489619 + 40320. error0 + 40320. exact0}
{Int[9], 623530.358939 - 362880. error0 - 362880. exact0}
```

$\{\text{Int}[10], -6.23530087111 \ 10^6 + 3.6288 \ 10^6 \ \text{error0} + 3.6288 \ 10^6 \ \text{exact0}\}$

$\{\text{Int}[11], 6.85883123005 \ 10^7 - 3.99168 \ 10^7 \ \text{error0} - 3.99168 \ 10^7 \ \text{exact0}\}$

$\{\text{Int}[12], -8.23059744888 \ 10^8 + 4.790016 \ 10^8 \ \text{error0} + 4.790016 \ 10^8 \ \text{exact0}\}$

G.8.a.i)   If the error0 is 0.0001, then how far off the correct values are the values of Int[0], Int[3], Int[7], and Int[12]?

If the error0 is 0.00000001, then how far off the correct values are the values of Int[0], Int[3], Int[7], and Int[12]?

G.8.a.ii) Describe the effect of the original error as the iteration progresses. Do small errors at the beginning result in larger or smaller errors at the end? Are small errors at the beginning anything to worry about?

G.8.b.i) Now it's time to turn the tables and make iteration work for you by iterating backward instead of forward. To see what this means, tackle the problem of seeing how to use backward iteration to get a very accurate approximate value of

$$\int_0^1 x^\pi e^x \, dx.$$

Try this:

Put

$$\text{Int}[k] = \int_0^1 x^{15-k+\pi} e^x \, dx.$$

How is Int[15] related to $\int_0^1 x^\pi e^x \, dx$?

G.8.b.ii) Use integration by parts to obtain a formula for $\text{Int}[k]$ in terms of $\text{Int}[k-1]$ for $k = 1, 2, \ldots, 15$.

G.8.b.iii) Agree that exact0 stands for the exact value of

$$\text{Int}[0] = \int_0^1 x^{15+\pi} e^x \, dx$$

and report on the error introduced to the calculation of

$$\text{Int}[15] = \int_0^1 x^\pi e^x \, dx$$

by incorporating an error into the value of Int[0]:

$$\text{Int}[0] = \text{exact0} + \text{error0}$$

and iterating.

G.8.b.iv) Why can you say at a glance that

$$0 < \text{Int}[0] = \int_0^1 x^{15+\pi} e^x \, dx < e?$$

If you cannot see this at a glance, then look at the following plot to see how the areas line up:

In[32]:=
```
Clear[x]
Plot[{0,x^(15 + Pi) E^x,E},
{x,0,1},PlotStyle->
{{Red},{Blue,Thickness[0.01]},{Red}},
AxesLabel->{"x",""},PlotRange->All];
```

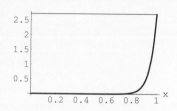

G.8.b.v)

If you use the value

$$0 = \text{Int}[0] = \text{exact0} + \text{error0},$$

then why is $|\text{error0}| < e < 3$?

G.8.b.vi)

If you use the value

$$0 = \text{Int}[0] = \text{exact0} + \text{error0},$$

then how many accurate decimals of $\text{Int}[15] = \int_0^1 x^\pi e^x \, dx$ are guaranteed, provided that no other calculational errors are made?

G.8.b.vii)

Give a more accurate estimate of the true value of $\int_0^1 x^\pi e^x \, dx$ than the estimate given by:

In[33]:=
```
Clear[x]
NIntegrate[x^Pi E^x,{x,0,1}]
```
Out[33]=
```
0.546881
```

Traditional Pat Integration Procedures for Special Situations

This lesson is not part of the mainline course. Those not interested in memorized rote procedures can safely skip it.

Basics

■ **B.1)** What do you mean by "pat integration procedures"? What is the role of pat integration procedures in a modern calculus course?

B.1.a) What does the word "pat" mean?

Answer: Go to the Webster's Dictionary as found on NeXT computers and look up the word *pat*. Here's what you get:

pat (adj) (1646), 1a: exactly suited to the purpose or occasion: APT. b: suspiciously appropriate: CONTRIVED.

2: learned, mastered, or memorized exactly.

3: FIRM, UNYIELDING.

4: reduced to a simple or mechanical form.

5: STANDARD, TRITE.

B.1.b) "You can tell the students not to have sex in the classroom and you can tell the students not to use computers or calculators in the classroom. They will probably comply with both edicts. But outside class . . .??"

Professor Thomas Tucker of Colgate University at a 1992 conference of math professors at Rensselear Polytechnic Institute

"I was helping a friend in the normal (read: old-fashioned, obsolete) section, and I worked a problem down to the integral and stopped there, satisfying myself that a computer could take it from there. My friend looked at me, stunned that I had not done the hardest part of the problem and considered myself finished. I was finished, for I had done the thinking behind the problem and didn't want to bother myself with the petty details of working through memorized procedures. That's what a computer's for."

1991 Calculus&*Mathematica* student Gerard Richardson

> Isn't it sort of silly for you to spend your time learning pat procedures to crank out integrals by hand when you've got a machine that is perfectly capable of handling them with little effort on your part? Don't you have better ways of occupying your time?

Answer: Back in the bad old days of calculus, a considerable part of the course was spent programming students with pat procedures to calculate integrals automatically by hand. *Mathematica* and other mathematics processors have forced a rethinking of this aspect of the course. As a result, what used to be the showcase topic of calculus has moved to the sidelines. More importantly, most serious integrals arising in real life don't fall within the realm of the pat procudures taught in the past. Still, you may have some curiosity about how to calculate integrals by hand because some of the hand procedures are exactly what are programmed into *Mathematica* for machine calculation of integrals. If you are interested in this, then go on with this lesson.

■ B.2) Pat procedure: Educated guessing—the method of undetermined coefficients

Look at *Mathematica*'s calculation of

$$\int_1^t x^2 \, e^x dx \; :$$

In[1]:=
```
Clear[x,t]; Integrate[x^2 E^(x),{x,0,t}]
```
Out[1]=
$$-2 + E^t \; (2 - 2 \, t + t^2)$$

To calculate $\int_1^t x^2 \, e^x \, dx$, you gotta come up with a function $f[x]$ with $f'[x] = x^2 \, e^x$.

$\rightarrow$ The formula for $f[x]$ should include $x^2 \, e^x$ because:

In[2]:=
```
D[x^2 E^x,x]
```

Out[2]=

$$2 E^x x + E^x x^2$$

→ and the formula for $f[x]$ should include $x\,e^x$ because:

In[3]:=
```
D[x E^x,x]
```

Out[3]=

$$E^x + E^x x$$

→ You can use the $x\,e^x$ term to cancel the $2\,x\,e^x$ term above.

Now you've got an extra e^x term to worry about; so you throw an extra e^x term into $f[x]$:

In[4]:=
```
Clear[a,b,c,x,prelimf]
prelimf[x_] = a x^2 E^x + b x E^x + c E^x
```

Out[4]=

$$c E^x + b E^x x + a E^x x^2$$

Now look at $D[\text{prelimf}[x], x]$:

In[5]:=
```
D[prelimf[x],x]
```

Out[5]=

$$b E^x + c E^x + 2 a E^x x + b E^x x + a E^x x^2$$

Collect the terms:

In[6]:=
```
Collect[D[prelimf[x],x],{E^x,x E^x,x^2 E^x}]
```

Out[6]=

$$E^x (b + c + 2 a x + b x + a x^2)$$

Because you are calculating $\int_1^t x^2\,e^x\,dx$, you want this to be $x^2\,e^x$, so you write

In[7]:=
```
eqe = (b + c == 0);
eqxe = (2 a + b == 0);
eqx2e = (a == 1);
```

And you solve:

In[8]:=
```
solvedcoefficients = Solve[{eqe,eqxe,eqx2e}]
```

Out[8]=
```
{{a -> 1, b -> -2, c -> 2}}
```

Substitute in to get the formula for $f[x]$:

In[9]:=
```
Clear[f]
f[x_] = prelimf[x]/.solvedcoefficients[[1]]
```

Out[9]=
$$2 E^x - 2 E^x x + E^x x^2$$

Most folks call this activity by the name "method of undetermined coefficients." Remembering that you wanted $f'[x] = x^2 e^x$, check $f'[x]$:

In[10]:=
```
f'[x]
```

Out[10]=
$$E^x x^2$$

Cool; now you can say for sure that $\int_1^t x^2 e^x \, dx$ is:

In[11]:=
```
Clear[t]; f[t] - f[1]
```

Out[11]=
$$-E + 2 E^t - 2 E^t t + E^t t^2$$

Check with your tireless mathematical servant:

In[12]:=
```
Expand[Integrate[x^2 E^x,{x,1,t}]]
```

Out[12]=
$$-E + 2 E^t - 2 E^t t + E^t t^2$$

Looks just fine.

B.2.a)
> Use the method of undetermined coefficients to calculate
> $$\int_0^t x \sin[x] \, dx.$$

Answer: To calculate $\int_0^t x \sin[x] \, dx$, you have to find a function $f[x]$ with $f'[x] = x \sin[x]$.

$\rightarrow$ The formula of $f[x]$ should include $x \cos[x]$ because:

In[13]:=
```
D[x Cos[x],x]
```

Out[13]=
```
Cos[x] - x Sin[x]
```

$\rightarrow$ Now you throw in enough extra terms to get rid of $\cos[x]$.

$$x \sin[x], \ x \cos[x], \ \cos[x], \ \sin[x]$$

ought to be enough.

In[14]:=
```
Clear[a,b,c,d,x,prelimf]
prelimf[x_] = a x Cos[x] + b x Sin[x] + c Cos[x] + d Sin[x]
```

Out[14]=
```
c Cos[x] + a x Cos[x] + d Sin[x] + b x Sin[x]
```

Now look at D[prelimf[x], x]:

In[15]:=
```
D[prelimf[x],x]
```

Out[15]=
```
a Cos[x] + d Cos[x] + b x Cos[x] + b Sin[x] - c Sin[x] - a x Sin[x]
```

Collect the terms:

In[16]:=
```
Collect[D[prelimf[x],x],{Cos[x],Sin[x]}]
```

Out[16]=
```
(a + d + b x) Cos[x] + (b - c - a x) Sin[x]
```

You want this to be $x \sin[x]$, so you write

In[17]:=
```
eqcos = a + d == 0;
eqxcos = b == 0;
eqsin = b - c  == 0;
eqxsin = -a == 1;
```

And you solve:

In[18]:=
```
solvedcoefficients = Solve[{eqcos,eqxcos,eqsin,eqxsin}]
```

Out[18]=
```
{{a -> -1, b -> 0, c -> 0, d -> 1}}
```

Substitute in to get the formula for $f[x]$:

In[19]:=
```
Clear[f]
f[x_] = prelimf[x]/.solvedcoefficients[[1]]
```

Out[19]=
```
-(x Cos[x]) + Sin[x]
```

Remembering that you wanted $f'[x] = x \sin[x]$, check $f'[x]$:

In[20]:=
```
f'[x]
```

Out[20]=
```
x Sin[x]
```

Perfecto; now you can say with poise and confidence that $\int_0^t x \sin[x] \, dx$ is:

In[21]:=
```
Clear[t]
f[t] - f[0]
```

Out[21]=

-(t Cos[t]) + Sin[t]

Check:

In[22]:=

Integrate[x Sin[x],{x,0,t}]

Out[22]=

-(t Cos[t]) + Sin[t]

Nailed it.

B.2.b) | Use the method of undetermined coefficients to calculate $\int_0^t e^{2x} \sin[3\,x]\,dx$.

Answer: To calculate $\int_0^t e^{2x} \sin[3\,x]\,dx$, the burden is on you to come up with a function $f[x]$ with $f'[x] = e^{2x} \sin[3\,x]$.

The formula of $f[x]$ should include $e^{2x} \cos[3\,x]$ because:

In[23]:=

D[E^(2 x) Cos[3 x],x]

Out[23]=

2 E$^{2\ x}$ Cos[3 x] - 3 E$^{2\ x}$ Sin[3 x]

Now you throw in enough extra terms to get rid of $e^{2x} \cos[3\,x]$. Throwing in $e^{2x} \sin[3\,x]$ ought to be enough.

In[24]:=

Clear[a,b,x,prelimf]
prelimf[x_] = a E^(2 x) Cos[3 x] + b E^(2 x) Sin[3 x]

Out[24]=

a E$^{2\ x}$ Cos[3 x] + b E$^{2\ x}$ Sin[3 x]

Now look at D[prelimf[x], x]:

In[25]:=

D[prelimf[x],x]

Out[25]=

2 a E$^{2\ x}$ Cos[3 x] + 3 b E$^{2\ x}$ Cos[3 x] - 3 a E$^{2\ x}$ Sin[3 x] + 2 b E$^{2\ x}$ Sin[3 x]

Collect the terms:

In[26]:=

Collect[D[prelimf[x],x],{Cos[3 x],Sin[3x]}]

Out[26]=

(2 a E$^{2\ x}$ + 3 b E$^{2\ x}$) Cos[3 x] + (-3 a E$^{2\ x}$ + 2 b E$^{2\ x}$) Sin[3 x]

Because you are calculating $\int_0^t e^{2t} \sin[3\,x]\,dx$, you want this to be $e^{2x} \sin[3\,x]$, so you write

In[27]:=
```
eqcos = (2 a + 3 b == 0);
eqsin = (-3 a + 2  b == 1);
```

And you solve:

In[28]:=
```
solvedcoefficients = Solve[{eqcos,eqsin}]
```

Out[28]=

$$\{\{a \to -(\frac{3}{13}),\ b \to \frac{2}{13}\}\}$$

Substitute in to get the formula for $f[x]$:

In[29]:=
```
Clear[f]
f[x_] = prelimf[x]/.solvedcoefficients[[1]]
```

Out[29]=

$$\frac{-3\ E^{2\ x}\ Cos[3\ x]}{13} + \frac{2\ E^{2\ x}\ Sin[3\ x]}{13}$$

Remembering that you wanted $f'[x] = e^{2x}\sin[3\,x]$, check $f'[x]$:

In[30]:=
```
f'[x]
```

Out[30]=

$$E^{2\ x}\ Sin[3\ x]$$

Impeccable; now you can say with aplomb that $\int_0^t e^{2x}\sin[3\,x]\,dx$ is:

In[31]:=
```
Clear[t]; f[t] - f[0]
```

Out[31]=

$$\frac{3}{13} - \frac{3\ E^{2\ t}\ Cos[3\ t]}{13} + \frac{2\ E^{2\ t}\ Sin[3\ t]}{13}$$

Check:

In[32]:=
```
Integrate[(E^(2 x)) Sin[3 x],{x,0,t}]
```

Out[32]=

$$\frac{3}{13} - \frac{3\ E^{2\ t}\ Cos[3\ t]}{13} + \frac{2\ E^{2\ t}\ Sin[3\ t]}{13}$$

Nailed it.

B.2.c)   Couldn't you have calculated the integrals above using integration by parts?

Answer: In principle, yes. In practice, the method of undetermined coefficients is usually easier to go with than integration by parts.

■ B.3) Pat procedure: More educated guessing—partial fractions for quotients of polynomials

Look at *Mathematica*'s evaluation of

$$\int_0^t \frac{4\,x+1}{(x+2)^3\,(x+1)^2}\,dx:$$

In[33]:=
```
Clear[x,t]
Integrate[(4 x + 1)/((x + 2)^3 ( x + 1)^2),{x,0,t}]
```
Out[33]=

$$-\left(\frac{71}{8}\right) + \frac{3}{1+t} + \frac{7}{2\,(2+t)^2} + \frac{10}{2+t} + 13\,\text{Log}[2] + 13\,\text{Log}[1+t] - 13\,\text{Log}[2+t]$$

To come up with this calculation by hand, take each of the simple factors of the denominator of

$$\frac{4\,x+1}{(x+2)^3\,(x+1)^2}$$

and find a convenient function whose derivative is a multiple of a corresponding factor. Try it:

$$\frac{1}{x+2} \longrightarrow \log[x+2]$$

$$\frac{1}{(x+2)^2} \longrightarrow \frac{1}{x+2}$$

$$\frac{1}{(x+2)^3} \longrightarrow \frac{1}{(x+2)^2}$$

$$\frac{1}{x+1} \longrightarrow \log[x+1]$$

$$\frac{1}{(x+1)^2} \longrightarrow \frac{1}{x+1}$$

In[34]:=
```
Clear[prelimf,x,a,b,c,d,e]
prelimf[x_] =
a/(1 + x) + b Log[1 + x] + c/(2 + x)^2 + d/(2 + x) + e Log[2 + x]
```
Out[34]=

$$\frac{a}{1+x} + \frac{c}{(2+x)^2} + \frac{d}{2+x} + b\,\text{Log}[1+x] + e\,\text{Log}[2+x]$$

You want
$$D[\text{prelimf}[x], x] = \frac{4x+1}{(x+2)^3 (x+1)^2}.$$

In[35]:=
```
birdnest = Together[D[prelimf[x],x]]
```

Out[35]=
```
(-8 a + 8 b - 2 c - 2 d + 4 e - 12 a x + 20 b x - 4 c x - 5 d x + 12 e x -
         2          2         2         2         3         3         3
  6 a x  + 18 b x  - 2 c x  - 4 d x  + 13 e x  - a x  + 7 b x  - d x  +
         3      4       4               2         3
  6 e x  + b x  + e x ) / ((1 + x)  (2 + x) )
```

This has the same denominator as
$$\frac{4x+1}{(x+2)^3 (x+1)^2}.$$

To make it equal to
$$\frac{4x+1}{(x+2)^3 (x+1)^2},$$

you find the *a*, *b*, *c*, *d*, and *e* that make the numerator equal to

$$4x+1.$$

The numerator of prelimf[x] is:

In[36]:=
```
num = Collect[Numerator[birdnest],{x}]
```

Out[36]=
```
-8 a + 8 b - 2 c - 2 d + 4 e + (-12 a + 20 b - 4 c - 5 d + 12 e) x +
                                               2                        3            4
  (-6 a + 18 b - 2 c - 4 d + 13 e) x  + (-a + 7 b - d + 6 e) x  + (b + e) x
```

Force this to be $4x+1$ by equating the coefficients of like powers of x:

In[37]:=
```
eq1 = (-8 a + 8 b - 2 c - 2 d + 4 e == 1);
eqx = (-12 a + 20 b - 4 c - 5 d + 12 e == 4);
eqxsquared = (-6 a + 18 b - 2 c - 4 d + 13 e == 0);
eqxcubed = (-a + 7 b - d + 6 e == 0);
eqxfourth = (b + e == 0);
solvedcoefficients = Solve[{eq1,eqx,eqxsquared,eqxcubed,eqxfourth}]
```

Out[37]=
```
                                          7
{{b -> 13, e -> -13, a -> 3, c -> -, d -> 10}}
                                          2
```

The function $f[x]$ with derivative
$$\frac{4x+1}{(x+2)^3 (x+1)^2}$$
is:

In[38]:=
```
Clear[f]
f[x_] = prelimf[x]/.solvedcoefficients[[1]]
```

Out[38]=

$$\frac{3}{1+x} + \frac{7}{2\ (2+x)^2} + \frac{10}{2+x} + 13\ \text{Log}[1+x] - 13\ \text{Log}[2+x]$$

Try it out in the expectation that

$$f'[x] = \frac{4\,x + 1}{(x+2)^3\,(x+1)^2}.$$

In[39]:=
```
f'[x]
```

Out[39]=

$$\frac{-3}{(1+x)^2} + \frac{13}{1+x} - \frac{7}{(2+x)^3} - \frac{10}{(2+x)^2} - \frac{13}{2+x}$$

Better go for a common denominator:

In[40]:=
```
Together[f'[x]]
```

Out[40]=

$$\frac{1+4\,x}{(1+x)^2\ (2+x)^3}$$

Hot Ziggety! Now you know that

$$\int_0^t \frac{4\,x+1}{(x+2)^3\,(x+1)^2}\ dx$$

is given by:

In[41]:=
```
Clear[t]; f[t] - f[0]
```

Out[41]=

$$-\left(\frac{71}{8}\right) + \frac{3}{1+t} + \frac{7}{2\ (2+t)^2} + \frac{10}{2+t} + 13\ \text{Log}[2] + 13\ \text{Log}[1+t] - 13\ \text{Log}[2+t]$$

Check:

In[42]:=
```
Integrate[(4 x + 1)/((x + 2)^3 ( x + 1)^2),{x,0,t}]
```

Out[42]=

$$-\left(\frac{71}{8}\right) + \frac{3}{1+t} + \frac{7}{2\ (2+t)^2} + \frac{10}{2+t} + 13\ \text{Log}[2] + 13\ \text{Log}[1+t] - 13\ \text{Log}[2+t]$$

Nailed it. Some folks call the method we used by the name "method of partial fractions."

Try the method of partial fractions to give a calculation of the integral

$$\int_0^t \frac{x^4}{(x-1)(x+3)^2}\, dx.$$

Answer: This time the polynomial in the numerator has a higher degree than the polynomial in the denominator. Respond by dividing the denominator into the numerator.

In[43]:=
```
PolynomialQuotient[x^4,(x - 1) (x + 3)^2,x]
```

Out[43]=
```
-5 + x
```

In[44]:=
```
PolynomialRemainder[x^4,(x - 1) (x + 3)^2,x]
```

Out[44]=
```
              2
-45 + 24 x + 22 x
```

This tells you that

$$\int_0^t \frac{x^4}{(x-1)(x+3)^2}\, dx = \int_0^t x - 5 + \frac{22\,x^2 + 24\,x - 45}{(x-1)(x+3)^2}\, dx$$

$$= \int_0^t (x - 5)\, dx + \int_0^t \frac{22\,x^2 + 24\,x - 45}{(x-1)(x+3)^2}\, dx$$

$$= \frac{t^2}{2} - 5\,t + \int_0^t \frac{22\,x^2 + 24\,x - 45}{(x-1)(x+3)^2}\, dx$$

Now work on

$$\int_0^t \frac{22\,x^2 + 24\,x - 45}{(x-1)(x+3)^2}\, dx.$$

To come up with this calculation by hand, take each of the simple factors of the denominator of

$$\frac{22\,x^2 + 24\,x - 45}{(x-1)(x+3)^2}$$

and find a convenient function whose derivative is a multiple of the corresponding factor. Try it:

$$\frac{1}{x - 1} \longrightarrow \log[x - 1]$$

$$\frac{1}{x + 3} \longrightarrow \log[x + 3]$$

$$\frac{1}{(x + 3)^2} \longrightarrow \frac{1}{x + 3}$$

In[45]:=
```
Clear[prelimf,x,a,b,c]
prelimf[x_] = a Log[x - 1] + b Log[x + 3] + c/(x + 3)
```

Out[45]=

$$\frac{c}{3 + x} + a \ Log[-1 + x] + b \ Log[3 + x]$$

You want

$$D[prelimf[x], x] = \frac{22\,x^2 + 24\,x - 45}{(x - 1)\,(x + 3)^2}.$$

In[46]:=
```
birdnest = Together[D[prelimf[x],x]]
```

Out[46]=

$$\frac{9\,a - 3\,b + c + 6\,a\,x + 2\,b\,x - c\,x + a\,x^2 + b\,x^2}{(-1 + x)\,(3 + x)^2}$$

This has the same denominator as

$$\frac{22\,x^2 + 24\,x - 45}{(x - 1)\,(x + 3)^2}.$$

To make it equal to

$$\frac{22\,x^2 + 24\,x - 45}{(x - 1)\,(x + 3)^2},$$

find the *a*, *b*, and *c* that make the numerator equal to $(22\,x^2 + 24\,x - 45)$. The numerator of prelimf[*x*] is:

In[47]:=
```
num = Collect[Numerator[birdnest],{x}]
```

Out[47]=

$$9\,a - 3\,b + c + (6\,a + 2\,b - c)\,x + (a + b)\,x^2$$

Force this to be $(22\,x^2 + 24\,x - 45)$ by equating the coefficients of like powers of x and solving:

In[48]:=
```
eq1 = (9 a - 3 b + c == -45);
eqx = (6a + 2 b - c == 24);
eqxsquared = (a + b == 22 );
solvedcoefficients = Solve[{eq1,eqx,eqxsquared}]
```

Out[48]=

$$\{\{a \to \frac{1}{16},\ b \to \frac{351}{16},\ c \to \frac{81}{4}\}\}$$

The function $f[x]$ with derivative

$$\frac{22\,x^2 + 24\,x - 45}{(x-1)\,(x+3)^2}$$

is:

In[49]:=
```
Clear[f]
f[x_] = prelimf[x]/.solvedcoefficients[[1]]
```

Out[49]=

$$\frac{81}{4\,(3 + x)} + \frac{\text{Log}[-1 + x]}{16} + \frac{351\,\text{Log}[3 + x]}{16}$$

Try it out in the expectation that

$$f'[x] = \frac{22\,x^2 + 24\,x - 45}{(x-1)\,(x+3)^2} :$$

In[50]:=
```
Together[f'[x]]
```

Out[50]=

$$\frac{-45 + 24\,x + 22\,x^2}{(-1 + x)\,(3 + x)^2}$$

Fat City! Now you know that

$$\int_0^t \frac{x^4}{(x-1)\,(x+3)^2}\,dx$$

is given by:

In[51]:=
```
Clear[t]
handCalculation = t^2/2 - 5 t + (f[t] - f[0])
```

Out[51]=

$$-\left(\frac{27}{4}\right) - \frac{I}{16}\,\text{Pi} - 5\,t + \frac{t^2}{2} + \frac{81}{4\,(3 + t)} - \frac{351\,\text{Log}[3]}{16} + \frac{\text{Log}[-1 + t]}{16} + \frac{351\,\text{Log}[3 + t]}{16}$$

Check:

In[52]:=
```
MathematicaCalculation = Integrate[(x^4)/((x - 1) (x + 3)^2),{x,0,t}]
```

Out[52]=

$$-5\,t + \frac{t^2}{2} + \frac{81}{4\,(3 + t)} - \frac{27\,(4 + 13\,\text{Log}[3])}{16} + \frac{\text{Log}[1 - t]}{16} + \frac{351\,\text{Log}[3 + t]}{16}$$

In[53]:=
```
Simplify[handCalculation - MathematicaCalculation]
```

Out[53]=
```
0
```

Got it.

■ B.4) Pat procedure: Trigonometric integrals

B.4.a) Here is *Mathematica*'s calculation of

$$\int_0^t \sin[x]^6 \, dx :$$

In[54]:=
```
Clear[x,t]; Integrate[Sin[x]^6,{x,0,t}]
```
Out[54]=

$$\frac{5\,t}{16} - \frac{15\,\mathrm{Sin}[2\,t]}{64} + \frac{3\,\mathrm{Sin}[4\,t]}{64} - \frac{\mathrm{Sin}[6\,t]}{192}$$

> Say how *Mathematica* might have come up with this result.

Answer: *Mathematica* probably applied some trig identities:

In[55]:=
```
betterintegrand = Expand[Sin[x]^6,Trig->True]
```
Out[55]=

$$\frac{5}{16} - \frac{15\,\mathrm{Cos}[2\,x]}{32} + \frac{3\,\mathrm{Cos}[4\,x]}{16} - \frac{\mathrm{Cos}[6\,x]}{32}$$

Now you can say that a function $f[x]$ whose derivative is $\sin[x]^6$ is:

In[56]:=
```
Clear[f]
f[x_] = 5 x/16 - 15 Sin[2 x]/((2)(32)) + 3 Sin[4 x]/((4)(16)) - Sin[6 x]/((6)(32))
```
Out[56]=

$$\frac{5\,x}{16} - \frac{15\,\mathrm{Sin}[2\,x]}{64} + \frac{3\,\mathrm{Sin}[4\,x]}{64} - \frac{\mathrm{Sin}[6\,x]}{192}$$

Check:

In[57]:=
```
{f'[x],betterintegrand}
```
Out[57]=

$$\{ \frac{5}{16} - \frac{15\,\mathrm{Cos}[2\,x]}{32} + \frac{3\,\mathrm{Cos}[4\,x]}{16} - \frac{\mathrm{Cos}[6\,x]}{32},$$

$$\frac{5}{16} - \frac{15\,\mathrm{Cos}[2\,x]}{32} + \frac{3\,\mathrm{Cos}[4\,x]}{16} - \frac{\mathrm{Cos}[6\,x]}{32} \}$$

Good. Now you can say with some measure of confidence that

$$\int_0^t \sin[x]^6 \, dx$$

is:

In[58]:=
  f[t] - f[0]

Out[58]=

$$\frac{5\ t}{16} - \frac{15\ \text{Sin}[2\ t]}{64} + \frac{3\ \text{Sin}[4\ t]}{64} - \frac{\text{Sin}[6\ t]}{192}$$

Check:

In[59]:=
  Integrate[Sin[x]∧6,{x,0,t}]

Out[59]=

$$\frac{5\ t}{16} - \frac{15\ \text{Sin}[2\ t]}{64} + \frac{3\ \text{Sin}[4\ t]}{64} - \frac{\text{Sin}[6\ t]}{192}$$

B.4.b) Here is *Mathematica*'s calculation of

$$\int_0^t \sin[a\,x]\ \cos[b\,x]\ dx$$

with $a \neq b$:

In[60]:=
  Clear[x,t,a,b]
  Integrate[Sin[a x] Cos[b x],{x,0,t}]

Out[60]=

$$-\left(\frac{a}{-a^2 + b^2}\right) + \frac{a\ \text{Cos}[a\ t]\ \text{Cos}[b\ t]}{-a^2 + b^2} + \frac{b\ \text{Sin}[a\ t]\ \text{Sin}[b\ t]}{-a^2 + b^2}$$

> Say how *Mathematica* might have come up with this result.

Answer: *Mathematica* probably applied some trig identities:

In[61]:=
  betterintegrand = Expand[Sin[a x] Cos[b x],Trig->True]

Out[61]=

$$\frac{\text{Sin}[a\ x - b\ x]}{2} + \frac{\text{Sin}[a\ x + b\ x]}{2}$$

Now you can say that a function $f[x]$ whose derivative is $\sin[a\,x]\ \cos[b\,x]$ is:

In[62]:=
  Clear[f]
  f[x_] = -Cos[(a - b) x]/((2) (a - b)) - Cos[(a + b) x]/((2) (a + b))

Out[62]=

$$\frac{-\text{Cos}[(a - b)\ x]}{2\ (a - b)} - \frac{\text{Cos}[(a + b)\ x]}{2\ (a + b)}$$

Check:

In[63]:=
```
{f'[x],betterintegrand}
```
Out[63]=

$$\{\frac{Sin[(a\ -\ b)\ x]}{2} + \frac{Sin[(a\ +\ b)\ x]}{2},$$

$$\frac{Sin[a\ x\ -\ b\ x]}{2} + \frac{Sin[a\ x\ +\ b\ x]}{2}\}$$

Good. Now you can say that

$$\int_0^t \sin[a\,x]\cos[b\,x]\,dx$$

is:

In[64]:=
```
patresult = ExpandAll[Together[f[t] - f[0]]]
```
Out[64]=

$$\frac{2\ a}{2\ a^2\ -\ 2\ b^2} - \frac{a\ Cos[a\ t\ -\ b\ t]}{2\ a^2\ -\ 2\ b^2} - \frac{b\ Cos[a\ t\ -\ b\ t]}{2\ a^2\ -\ 2\ b^2} -$$

$$\frac{a\ Cos[a\ t\ +\ b\ t]}{2\ a^2\ -\ 2\ b^2} + \frac{b\ Cos[a\ t\ +\ b\ t]}{2\ a^2\ -\ 2\ b^2}$$

Check:

In[65]:=
```
MathematicaCalculation = Expand[Integrate[Sin[a x] Cos[b x],{x,0,t}],Trig->True]
```
Out[65]=

$$-(\frac{a}{-a^2\ +\ b^2}) + \frac{a\ Cos[a\ t\ -\ b\ t]}{2\ (-a^2\ +\ b^2)} + \frac{b\ Cos[a\ t\ -\ b\ t]}{2\ (-a^2\ +\ b^2)} +$$

$$\frac{a\ Cos[a\ t\ +\ b\ t]}{2\ (-a^2\ +\ b^2)} - \frac{b\ Cos[a\ t\ +\ b\ t]}{2\ (-a^2\ +\ b^2)}$$

In[66]:=
```
Simplify[patresult - MathematicaCalculation]
```
Out[66]=
```
0
```

OK.

■ B.5) Pat procedure: Wild card substitution

B.5.a) Here is *Mathematica*'s calculation of

$$\int_0^1 \left(1 - x^2\right)^{3/2}\,dx :$$

In[67]:=
```
Clear[x]
MathematicaCalculation = Integrate[(1 - x^2)^(3/2),{x,0,1}]
```

Out[67]=

$$\frac{3 \text{ Pi}}{16}$$

Use a trig substitution to explain where this result comes from.

Answer: The integral absorbing your attention is

$$\int_0^1 \left(1 - x^2\right)^{3/2} \, dx.$$

There is a definite geek in this integral; it's the term $\left(1 - x^2\right)^{3/2}$.
Squash the geek by playing your wild card and setting $x = \sin[t]$.
This substitution gives the pairings

$$\left(1 - x^2\right)^{3/2} \longleftrightarrow \left(1 - \sin[t]^2\right)^{3/2} = \left(\cos[t]^2\right)^{3/2} = \cos[t]^3;$$

$$dx \longleftrightarrow dt;$$

$$\int_0^1 \longleftrightarrow \int_0^{\pi/2}$$

because $t = 0$ when $x = 0$ and $t = \pi/2$ when $x = 1$. This tells you that

$$\int_0^1 \left(1 - x^2\right)^{3/2} \, dx = \int_0^{\pi/2} \cos[t]^3 \, \cos[t] \, dt = \int_0^{\pi/2} \cos[t]^4 \, dt.$$

Now apply some trig identities:

In[68]:=
```
Clear[t]; Expand[Cos[t]^4,Trig->True]
```

Out[68]=

$$\frac{3}{8} + \frac{\text{Cos}[2\text{ t}]}{2} + \frac{\text{Cos}[4\text{ t}]}{8}$$

This tells you that

$$\int_0^1 \left(1 - x^2\right)^{3/2} \, dx = \int_0^{\pi/2} \cos[t]^4 \, dt = \int_0^{\pi/2} \left(\frac{3}{8} + \frac{\cos[2\,t]}{2} + \frac{\cos[4\,t]}{8}\right) \, dt.$$

Now all you've got to do is come up with a function $f[t]$ with

$$f'[t] = \frac{3}{8} + \frac{\cos[2\,t]}{2} + \frac{\cos[4\,t]}{8}.$$

Because $D[\sin[a\,t], t] = a \, \cos[a\,t]$, one such $f[t]$ is:

In[69]:=
```
Clear[f,t]; f[t_] = 3 t/8 + Sin[2 t]/4 + Sin[4 t]/32
```
Out[69]=

$$\frac{3 t}{8} + \frac{Sin[2 t]}{4} + \frac{Sin[4 t]}{32}$$

Check:

In[70]:=
```
f'[t]
```
Out[70]=

$$\frac{3}{8} + \frac{Cos[2 t]}{2} + \frac{Cos[4 t]}{8}$$

Good. Now you've explained why

$$\int_0^1 \left(1 - x^2\right)^{3/2} dx = \int_0^{\pi/2} f'[t]\, dt$$

is given by:

In[71]:=
```
f[Pi/2] - f[0]
```
Out[71]=

$$\frac{3 Pi}{16}$$

Check:

In[72]:=
```
MathematicaCalculation
```
Out[72]=

$$\frac{3 Pi}{16}$$

Nailed it. Squashing the geeks can be perverse fun.

B.5.b.i) The hyperbolic functions $\sinh[x]$ and $\cosh[x]$ are defined by

$$\sinh[x] = \frac{e^x - e^{-x}}{2}$$

and

$$\cosh[x] = \frac{e^x + e^{-x}}{2}.$$

Note the similarities with $\sin[x]$ and $\cos[x]$:

In[73]:=
```
Clear[x]; {D[Sinh[x],x],D[Sin[x],x]}
```
Out[73]=
```
{Cosh[x], Cos[x]}
```

This tells you $D[\sinh[x], x] = \cosh[x]$ just like $D[\sin[x], x] = \cos[x]$.

In[74]:=
```
Clear[x]; {D[Cosh[x],x],D[Cos[x],x]}
```

Out[74]=
```
{Sinh[x], -Sin[x]}
```

This tells you $D[\cosh[x], x] = \sinh[x]$ almost like $D[\cos[x], x] = -\sin[x]$.

The identity connecting $\sinh[x]$ with $\cosh[x]$ is almost the same as the identity connecting $\sin[x]$ and $\cos[x]$:

In[75]:=
```
Clear[x]
Expand[1 + Sinh[x]^2,Trig->True] == Expand[Cosh[x]^2,Trig->True]
```

Out[75]=
```
True
```

In[76]:=
```
Clear[t]
Expand[1 - Sin[x]^2,Trig->True] == Expand[Cos[x]^2,Trig->True]
```

Out[76]=
```
True
```

The result: $1 + \sinh[x]^2 = \cosh[x]$ as opposed to $1 - \sin[x]^2 = \cos[x]^2$.

Why are these identities useful in hand calculation of certain integrals?

Answer: You can use $x = a \sin[t]$ to condense $a^2 - x^2$ into a single term via
$$a^2 - x^2 = a^2 - a^2 \sin[t]^2 = a^2 \left(1 - \sin[t]^2\right) = a^2 \cos[t]^2.$$

This is what was done in part B.5.b.a) above.

You can use $x = a \sinh[t]$ to condense $a^2 + x^2$ into a single term via
$$a^2 + x^2 = a^2 + a^2 \sinh[t]^2 = a^2 \left(1 + \sinh[t]^2\right) = a^2 \cosh[t]^2.$$

This is what will be done in part B.5.b.ii) immediately below. Watch for it.

B.5.b.ii)   Here is *Mathematica*'s calculation of
$$\int_0^2 \left(x^2 + 4\right)^{3/2} dx :$$

In[77]:=
```
Clear[x]
MathematicaCalculation = Integrate[(x^2 + 4)^(3/2),{x,0,2}]
```

Out[77]=
```
     3/2
  7 2    + 6 ArcSinh[1]
```

Use a wild card substitution to explain where this result comes from.

Answer: The integral you're looking at is $\int_0^2 \left(x^2 + 4\right)^{3/2} dx$. There is a definite geek in this integral; it's the term $\left(x^2 + 4\right)^{3/2}$. The idea is to make a wild card substitution that squashes this geek by simplifying $\left(x^2 + 4\right)^{3/2}$ into a single term. Here are two identities from part B.5.b.i) above:

$$1 + \sinh[x]^2 = \cosh[x]$$

and

$$1 - \sin[x]^2 = \cos[x]^2.$$

The first is the one you want because it says

$$4 + 4 \sinh[t]^2 = 4 \cosh[t]^2$$

and so

$$\left(4 + 4 \sinh[t]^2\right)^{3/2} = \left(4 \cosh[t]^2\right)^{3/2} = 4^{3/2} \cosh[t]^3.$$

This tells you that you can squash the geek $\left(x^2 + 4\right)^{3/2}$ in $\int_0^2 \left(x^2 + 4\right)^{3/2} dx$ by making the wild card substitution

$$x[t] = 2 \sinh[t].$$

Because $x'[t] = 2 \cosh[t]$, this gives you the pairings

$$\left(1 + x^2\right)^{3/2} \longleftrightarrow \left(4 \cosh[t]^2\right)^{3/2} = 4^{3/2} \cosh[t]^3;$$

$$dx \longleftrightarrow dt;$$

$$\int_0^2 \longleftrightarrow \int_0^{\text{ArcSinh}[1]}$$

because when $t = \text{ArcSinh}[1]$, then

$$x[t] = 2 \sinh[\text{ArcSinh}[1]] = 2$$

and when $t = 0$, then

$$x[t] = 2 \sinh[0] = \frac{e^0 - e^{-0}}{2} = 0.$$

The upshot:

$$\int_0^2 \left(x^2 + 4\right)^{3/2} dx = \int_0^{\text{ArcSinh}[1]} 4^{3/2} \cosh[t]^3 \, 2 \cosh[t] \, dt$$

$$= \int_0^{\text{ArcSinh}[1]} (2) \, 4^{3/2} \cosh[t]^4 \, dt.$$

To get your hands on this last integral, apply some identities:

In[78]:=

```
Clear[t]
Expand[(2) 4^(3/2) Cosh[t]^4,Trig->True]
```

Out[78]=

```
6 + 8 Cosh[2 t] + 2 Cosh[4 t]
```

This tells you that

$$\int_0^2 \left(x^2 + 4\right)^{3/2} dx = \int_0^{\text{ArcSinh}[1]} (2)\ 4^{3/2}\ \cosh[t]^4\ dt$$

$$= \int_0^{\text{ArcSinh}[1]} \left(6 + 8\ \cosh[2\,t] + 2\ \cosh[4\,t]\right)\ dt.$$

Now all you've got to do is come up with a function $f[t]$ with

$$f'[t] = 6 + 8\ \cosh[2\,t] + 2\ \cosh[4\,t].$$

Because $D[\sinh[a\,t], t] = a\ \cosh[a\,t]$, one such $f[t]$ is:

In[79]:=
```
Clear[f,t]
f[t_] = 6 t + 8 Sinh[2 t]/2 + 2 Sinh[4 t]/4
```
Out[79]=

$$6\ t + 4\ \text{Sinh}[2\ t] + \frac{\text{Sinh}[4\ t]}{2}$$

Check:

In[80]:=
```
f'[t]
```
Out[80]=
```
6 + 8 Cosh[2 t] + 2 Cosh[4 t]
```

Good. Now you've explained why $\int_0^2 \left(x^2 + 4\right)^{3/2} dx = \int_0^{\text{ArcSinh}[1]} f'[t]\ dt$ is given by:

In[81]:=
```
handanswer = f[ArcSinh[1]] - f[0]
```
Out[81]=

$$6\ \text{ArcSinh}[1] + 4\ \text{Sinh}[2\ \text{ArcSinh}[1]] + \frac{\text{Sinh}[4\ \text{ArcSinh}[1]]}{2}$$

Check:

In[82]:=
```
MathematicaCalculation
```
Out[82]=

$$7\ 2^{3/2} + 6\ \text{ArcSinh}[1]$$

Uh-oh; not the same formula. Check the numerical values:

In[83]:=
```
{N[handanswer],N[MathematicaCalculation]}
```
Out[83]=
```
{25.0872, 25.0872}
```

It checks. There was never a doubt.

B.5.c)   Calculate
$$\int_1^8 \frac{1}{1 + x^{1/3}}\, dx.$$

Answer: There's a smelly geek in $\int_1^8 1/\left(1 + x^{1/3}\right)\, dx$ and it's the $x^{1/3}$ term. Squash the geek by transforming this integral with the substitution $t^3 = x$. This gives the pairings

$$\frac{1}{1 + x^{1/3}} \longleftrightarrow \frac{1}{1 + t};$$
$$dx \longleftrightarrow 3\,t^2\, dt;$$
$$\int_1^8 \longleftrightarrow \int_1^2$$

because $t = 1$ when $x = 1$ and $t = 2$ when $x = 8$. So

$$\int_1^8 \frac{1}{1 + x^{1/3}}\, dx = \int_1^2 \frac{3\,t^2}{1 + t}\, dt.$$

To handle $\int_1^2 3\,t^2/(1 + t)\, dt$, apply some algebra to $3\,t^2/(1 + t)$:

In[84]:=
```
Clear[t]; Apart[3 (t^2)/(1 + t)]
```

Out[84]=
```
            3
-3 + 3 t + ─────
           1 + t
```

This tells you that

$$\int_1^8 \frac{1}{1 + x^{1/3}}\, dx = \int_1^2 \frac{3\,t^2}{1 + t}\, dt = \int_1^2 -3 + 3\,t + \frac{3}{1 + t}\, dt.$$

To calculate this, go for a function $f[t]$ with

$$f'[t] = -3 + 3\,t + \frac{3}{1 + t}.$$

Here's one:

In[85]:=
```
Clear[f]
f[t_] = -3 t + 3 (t^2)/2 + 3 Log[1 + t];
f'[t]
```

Out[85]=
```
            3
-3 + 3 t + ─────
           1 + t
```

Using this $f[t]$, you find

$$\int_1^8 \frac{1}{1 + x^{1/3}}\, dx = \int_1^2 -3 + 3\,t + \frac{3}{1 + t}\, dt$$

is given by:

In[86]:=
```
f[2] - f[1]
```

Out[86]=

$$\frac{3}{2} - 3\ \text{Log}[2] + 3\ \text{Log}[3]$$

Check:

In[87]:=
```
Integrate[1/(1 + x^(1/3)),{x,1,8}]
```

Out[87]=

$$\frac{3}{2} - 3\ \text{Log}[2] + 3\ \text{Log}[3]$$

■ B.6) Pat procedure: Integration by parts

This problem appears only in the electronic version.

Tutorials

■ T.1) Traditional pat procedures: A frank discussion

T.1.a)

What do the following integrals have in common?

$$\int_0^1 e^x \log[x]\,dx \qquad \int_0^\pi e^{\sin[x]}\,dx \qquad \int_0^1 x\tan[x]\,dx$$

$$\int_0^\infty \frac{e^{-x}}{1+x}\,dx \qquad \int_1^4 \sin[x^2]\,dx \qquad \int_1^3 \frac{\sin[x]}{x}\,dx$$

Answer: As innocent as these integrals look, none of the traditional pat procedures will handle any of them. Traditional calculus courses seldom mentioned that there are many integrals that can't be handled by the traditional pat procedures. *Mathematica* can calculate some of them symbolically. NIntegrate will handle them all. Get the message?

T.1.b)

Will you be hurt in advanced mathematics, engineering, and science courses if you are not able to handle the traditional pat procedures of integration?

Answer: Strangely enough, the answer is a definite no. You will be hurt badly if you can't handle transformations of integrals, and you will look slightly illiterate if you say you've never heard of integration by parts. The other traditional pat procedures are unlikely to come up outside a traditional calculus classroom.

Reason: The vast majority of integrals that come up outside the traditional calculus classroom can't be handled by the traditional pat procedures. One reason for this is that nearly all scientific endeavors beyond calculus itself involve the exponential function e^x. When you review the traditional pat procedures, you will note that few of them deal with calculation of integrals involving e^x. This observation leads to the conclusion that memorizing and executing the traditional pat integration procedures is probably not a good use of your time. This observation also leads to the conclusion that if you don't expend much effort on the traditional pat procedures of integration, you will not be hurt in the future.

The only people gunning for you might be frustrated traditional math teachers who believe that calculus consists chiefly of memorized rote procedures to be done by hand without machine help. These mathematical moralists will disappear as time goes on. In fact, their ranks should decay exponentially.

■ T.2) Integrals to try

This problem appears only in the electronic version.

■ T.3) Complex numbers and partial fractions

T.3.a)   Explain how to calculate an integral like

$$\int_0^t \frac{2\,x - 6}{x^3 - 5\,x^2 + 7\,x + 13}\,dx.$$

Answer: There are several ways of going about calculating this integral. Because you know something about complex numbers, complex numbers is the way to go.

Feed in the integrand:

In[1]:=
```
Clear[integrand,x]
integrand[x_] = (2 x - 6)/(x^3 - 5 x^2 + 7 x + 13)
```

Out[1]=
$$\frac{-6 + 2\ x}{13 + 7\ x - 5\ x^2 + x^3}$$

Find the complex roots of the denominator:

In[2]:=
```
Solve[Denominator[integrand[x]] == 0,x]
```

Out[2]=
```
{{x -> -1}, {x -> 3 - 2 I}, {x -> 3 + 2 I}}
```

To calculate this integral, take each of the simple factors of the denominator of integrand[x] and find a convenient function whose derivative is a multiple of the corresponding factor.

$$\frac{1}{x+1} \longrightarrow \log[x+1]$$

$$\frac{1}{x-(3-2i)} \longrightarrow \log[x-(3-2i)]$$

$$\frac{1}{x-(3+2i)} \longrightarrow \log[x-(3+2i)]$$

In[3]:=
```
Clear[prelimf,x,a,b,c]
prelimf[x_] = a Log[x + 1] + b Log[x - (3 - 2 I)] + c Log[x - (3 + 2 I)]
```

Out[3]=
```
c Log[-3 - 2 I + x] + b Log[-3 + 2 I + x] + a Log[1 + x]
```

Take the derivative of prelimf[x] and organize it:

In[4]:=
```
prelimfprime = Together[ExpandAll[Together[D[prelimf[x],x]]]]
```

Out[4]=
```
(13 a + (-3 - 2 I) b + (-3 + 2 I) c - 6 a x +

                                         2     2     2
    (-2 - 2 I) b x + (-2 + 2 I) c x + a x  + b x  + c x ) /

                 2    3
    (13 + 7 x - 5 x  + x )
```

The denominator matches the denominator of the integrand

$$\frac{2x-6}{x^3 - 5x^2 + 7x + 13}.$$

Look at the numerator:

In[5]:=
```
numer = Collect[Numerator[prelimfprime],{1,x,x^2}]
```

Out[5]=
```
13 a + (-3 - 2 I) b + (-3 + 2 I) c +

                                                        2
    (-6 a + (-2 - 2 I) b + (-2 + 2 I) c) x + (a + b + c) x
```

You want D[prelimf[x], x] = integrand[x]. To make the numerators match, you want:

In[6]:=
```
eqn1 = 13 a + (-3 - 2 I) b + (-3 + 2 I) c == -6;
eqn2 = -6 a + (-2 - 2 I) b + (-2 + 2 I) c == 2;
eqn3 = a + b + c == 0;
sols = Solve[{eqn1,eqn2,eqn3}]
```

Out[6]=
```
              2          1    I        1    I
    {{a -> -(-), b -> - + --, c -> - - --}}
              5          5   10        5   10
```

The $f[x]$ you want is:

In[7]:=
```
Clear[f,x]; f[x_] = prelimf[x]/.sols[[1]]
```
Out[7]=

$$\left(\frac{1}{5} - \frac{I}{10}\right) \text{Log}[-3 - 2\,I + x] + \left(\frac{1}{5} + \frac{I}{10}\right) \text{Log}[-3 + 2\,I + x] - \frac{2\,\text{Log}[1 + x]}{5}$$

Going with this function, you find that

$$\int_0^t \frac{2\,x - 6}{x^3 - 5\,x^2 + 7\,x + 13}\,dx$$

is given by:

In[8]:=
```
f[t] - f[0]
```
Out[8]=

$$\left(-\left(\frac{1}{5}\right) + \frac{I}{10}\right) \text{Log}[-3 - 2\,I] + \left(-\left(\frac{1}{5}\right) - \frac{I}{10}\right) \text{Log}[-3 + 2\,I] +$$

$$\left(\frac{1}{5} - \frac{I}{10}\right) \text{Log}[-3 - 2\,I + t] + \left(\frac{1}{5} + \frac{I}{10}\right) \text{Log}[-3 + 2\,I + t] - \frac{2\,\text{Log}[1 + t]}{5}$$

Weird looking; compare to the *Mathematica* calculation:

In[9]:=
```
Mathematicaf[t_] = Integrate[integrand[x],{x,0,t}]
```
Out[9]=

$$\frac{\text{ArcTan}[8] - \text{Log}[13]}{5} + \frac{\text{ArcTan}\left[\frac{2\,(-4 + t)}{1 + t}\right] - 2\,\text{Log}[1 + t] + \text{Log}[13 - 6\,t + t^2]}{5}$$

The formulas don't look the same. To see whether they agree, plot both together:

In[10]:=
```
Plot[{Mathematicaf[t],
f[t] - f[0]},{t,0,4},
PlotStyle->{{Blue,Thickness[0.01]},{Red}},
AspectRatio->1/GoldenRatio];
```

Cohabiting all the way. This is a case in which your answer doesn't have the same formula as the formula for the answer in the back of the book. The plot gives you the comfort of knowing that even though the result of your calculation does not look like the result of the *Mathematica* calculation, both results are the same. Done.

T.3.b) This problem appears only in the electronic version.

Give It a Try

This lesson is not part of the mainline course. Those not interested in memorized rote procedures can safely skip this lesson.

G.1)

Use a pat procedure to calculate

$$\int_0^t \cos[x]^4 \sin[x]^2 \, dx.$$

G.2)

Calculate

$$\int_0^t \sin[x]^3 \cos[x] \, dx.$$

G.3)

Use a pat procedure to calculate

$$\int_0^t x^3 \sin[x] \, dx.$$

G.4)

Use a pat procedure to calculate

$$\int_0^t x^8 e^{-x} \, dx.$$

G.5)

Use a pat procedure to calculate

$$\int_2^3 \frac{1}{x^2 - 4x - 5} \, dx.$$

G.6)

Use a pat procedure to calculate

$$\int_2^3 \frac{1}{x^2 - 4x + 5} \, dx.$$

G.7)

Use a pat procedure to calculate

$$\int_0^t \frac{5x^2 + x - 3}{x^4 + 10x^3 + 36x^2 + 56x + 32} \, dx.$$

G.8)   Use a pat procedure to calculate

$$\int_0^{1.6} x^3 \left(9 - x^2\right)^{3/2} dx.$$

G.9)   Use a pat procedure to calculate

$$\int_0^6 x^3 \left(9 + x^2\right)^{3/2} dx.$$

G.10)   Use a pat procedure to calculate

$$\int_3^{12} x^3 \left(x^2 - 9\right)^{3/2} dx.$$

G.11)   Use a pat procedure to calculate

$$\int_1^{27} \frac{1}{1 + x^{1/3}} dx.$$

G.12)   Use a pat procedure to calculate

$$\int_0^t \text{ArcTan}[x] \, dx.$$

G.13)   Use a pat procedure to calculate

$$\int_0^t \text{SinIntegral}[x] \, dx.$$

G.14)   Calculate

$$\int_0^t \frac{e^{-3x}}{1 + e^{-3x}} dx.$$

Integrals: Measurements of Accumulated Growth

2.01 Integrals for Measuring Area Literacy Sheet

L.1) Here's a plot of $f[x] = 4\,x\,e^{-x^2}$:

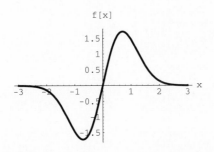

Look at the plot and then use the words *positive*, *negative*, or *zero* to complete the blanks:

$\displaystyle\int_{-3}^{3} f[x]\,dx$ is _____ .

$\displaystyle\int_{0}^{3} f[x]\,dx$ is _____ .

$\displaystyle\int_{-3}^{0} f[x]\,dx$ is _____ .

$\displaystyle\int_{-1}^{2} f[x]\,dx$ is _____ .

L.2) Here's a plot of $f[x] = 2\,x\,e^{0.5x}$:

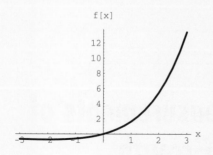

Look at the plot and then use the words *positive*, *negative*, or *zero* to complete the blanks:

$\displaystyle\int_{-3}^{3} f[x]\,dx$ is _____.

$\displaystyle\int_{0}^{3} f[x]\,dx$ is _____.

$\displaystyle\int_{-3}^{0} f[x]\,dx$ is _____.

$\displaystyle\int_{-1}^{2} f[x]\,dx$ is _____.

L.3) Here's a plot of $f[x] = 2\,\sin[x]^2$:

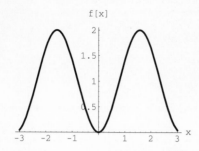

Look at the plot and then use the words *positive*, *negative*, or *zero* to complete the blanks:

$\displaystyle\int_{-3}^{3} f[x]\,dx$ is _____.

$\displaystyle\int_{0}^{3} f[x]\,dx$ is _____.

$$\int_{-3}^{0} f[x]\, dx \text{ is } \underline{\hspace{3cm}}.$$

$$\int_{-1}^{2} f[x]\, dx \text{ is } \underline{\hspace{3cm}}.$$

L.4) How do you know in advance with no calculation that:

$$\int_{-2}^{2} x^3\, dx = 0, \quad \int_{-1}^{2} x^3\, dx > 0, \quad \text{and} \quad \int_{-2}^{1} x^3\, dx < 0?$$

L.5) Use the axes below to give a hand sketch of the plot of

$$f[x] = \sin[x] \qquad \text{for } 0 \le x \le 2\pi.$$

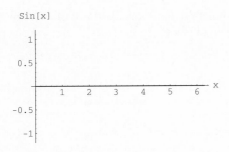

How does your plot signal that:

$$\int_{0}^{2\pi} \sin[x]\, dx = 0, \quad \int_{0}^{\pi} \sin[x]\, dx > 0, \quad \text{and} \quad \int_{\pi}^{2\pi} \sin[x]\, dx < 0?$$

L.6) Use the axes below to give a hand sketch of the plot of

$$f[x] = \cos[x] \qquad \text{for } 0 \le x \le \pi.$$

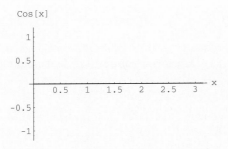

How does your plot signal that:

$$\int_{0}^{\pi} \cos[x]\, dx = 0, \quad \int_{0}^{\pi/2} \cos[x]\, dx > 0, \quad \text{and} \quad \int_{\pi/2}^{\pi} \cos[x]\, dx < 0?$$

L.7) Here is a blank plot:

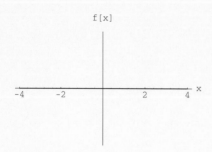

Pencil in the graph of any function $f[x]$ you like that takes some positive values and some negative values as x advances from -4 to 4. Shade your plot to indicate what

$$\int_{-4}^{4} f[x]\, dx$$

measures.

L.8) You are given numbers a and b with $a < b$. You are also given a function $f[x]$. Does the area measurement

$$\int_{a}^{b} f[x]\, dx$$

turn out to be a number or a function?

L.9) You are given numbers a and b with $a < b$. You are also given a function $f[x]$ that is always positive as x advances from a to b. Is it possible that

$$\int_{a}^{b} f[x]\, dx$$

turns out to be negative? Why?

L.10) You are given new numbers a and b with $a < b$. You are also given a new function $f[x]$ that is always negative as x advances from a to b. Is it possible that

$$\int_{a}^{b} f[x]\, dx$$

turns out to be positive? Why?

L.11) You are given new numbers a and b with $a < b$. You are also given two new functions $f[x]$ and $g[x]$ that are both always positive as x advances from a to b. If $f[x] > g[x]$ for all the x's between a and b, then why is it automatic that

$$\int_{a}^{b} f[x]\, dx > \int_{a}^{b} g[x]\, dx?$$

L.12) Calculate

$$\int_0^4 x \, dx.$$

L.13) When you plot

$$f[x] = \sqrt{4 - x^2} \qquad \text{for } -2 \le x \le 2,$$

you get the top of the circle of radius 2 centered at $\{0, 0\}$:

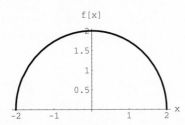

The area enclosed by a circle of radius r measures out to πr^2 square units. Use this fact to calculate

$$\int_{-2}^2 f[x] \, dx \qquad \text{and} \qquad \int_0^2 f[x] \, dx.$$

L.14) When you plot

$$f[x] = -\sqrt{9 - x^2} \qquad \text{for } -3 \le x \le 3,$$

you get the bottom of the circle of radius 3 centered at $\{0, 0\}$:

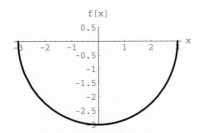

The area enclosed by a circle of radius r measures out to πr^2 square units. Use this fact to calculate

$$\int_{-3}^3 f[x] \, dx \qquad \text{and} \qquad \int_0^3 f[x] \, dx.$$

L.15) Explain the idea behind the formula

$$\int_a^b f[x] \, dx = \int_a^c f[x] \, dx + \int_c^b f[x] \, dx$$

for any number c with $a < c < b$.

L.16) Explain the idea behind the formula

$$\int_a^b K f[x]\, dx = K \int_a^b f[x]\, dx$$

for any number K.

L.17) Explain the idea behind the formula

$$\int_a^b f[x]\, dx = \int_a^b f[t]\, dt.$$

L.18) The graph of a function $f[x]$ consists of two straight line segments. The first segment connects $\{0,1\}$ and $\{2,2\}$ and the second segment connects $\{2,2\}$ and $\{3,0\}$.

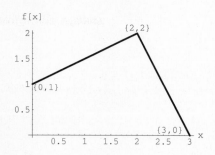

Measure some areas to come up with a calculation of

$$\int_0^3 f[x]\, dx.$$

L.19) What's the idea behind integration by trapezoids?

L.20) Go with

$$f[x] = 2 + 0.05\, x\, \sin[3\, x^2]$$

and look at this embellished plot of $f[x]$ for $1 \le x \le 4$:

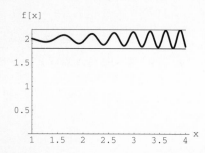

Explain how the plot tells you that when you estimate

$$\int_1^3 f[x]\, dx = 8,$$

then your estimate is not bad at all.

2.02 Breaking the Code of the Integral: The Fundamental Formula Literacy Sheet

L.1) When you calculate

$$\int_0^t 4\,\sin[x^3]\,dx$$

numerically for various values of t, you get different numbers depending on what t you go with:

In[1]:=
```
Clear[x]; t = 0.5;
NIntegrate[4 Sin[x^3],{x,0,t},AccuracyGoal->2]
```

Out[1]=
```
0.0624349
```

In[2]:=
```
t = 1.0;
NIntegrate[4 Sin[x^3],{x,0,t},AccuracyGoal->2]
```

Out[2]=
```
0.935381
```

In[3]:=
```
t = 1.5;
NIntegrate[4 Sin[x^3],{x,0,t},AccuracyGoal->2]
```

Out[3]=
```
2.34753
```

The upshot: When you put

$$f[t] = \int_0^t 4\,\sin[x^3]\,dx,$$

you can plot $f[t]$ as a function of t:

In[4]:=
```
Clear[f,t,x]
f[t_] := NIntegrate[4 Sin[x^3],
{x,0,t},AccuracyGoal->2];
Plot[f[t],{t,0,2},
PlotStyle->Thickness[0.01],
AxesLabel->{"t","f[t]"}];
```

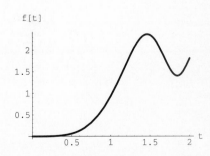

Even though you don't have a clean formula for $f[t]$, you can give a clean formula for $f'[t]$. Do it.

L.2) When you start with $f[x] = \sin[x]$ and remember that $\sin[0] = 0$ and $\sin[\pi/2] = 1$, you can easily write down the value of

$$\int_0^{\pi/2} f'[x]\, dx = \int_0^{\pi/2} \cos[x]\, dx.$$

Do it.

L.3) When you feed a formula for a function $g[x]$ into *Mathematica* and ask *Mathematica* to calculate

$$\int_a^t g[x]\, dx,$$

Mathematica looks for a function $f[t]$ with $f'[t] = g[t]$ and then *Mathematica* reports

$$\int_a^t g[x]\, dx = f[t] - f[a].$$

Once a function $f[t]$ with $f'[t] = g[t]$ is found, then the reported answer is guaranteed to be correct because the fundamental formula says that

$$\int_a^t g[x]\, dx = \int_a^t f'[x]\, dx = f[t] - f[a].$$

When you calculate integrals, you do the same thing. Calculate the following integrals without the help of any machine.

a) $\displaystyle\int_0^2 x\, dx$

b) $\displaystyle\int_0^2 x^2\, dx$

c) $\displaystyle\int_0^2 \left(x^3 + 3x + 1\right)\, dx$

d) $\displaystyle\int_a^t e^x\, dx$

e) $\displaystyle\int_a^t e^{2x}\, dx$

f) $\displaystyle\int_a^t e^{-2x}\, dx$

g) $\displaystyle\int_0^t \cos[x]\, dx$

h) $\displaystyle\int_0^t \cos[2x]\, dx$

i) $\displaystyle\int_0^t \sin[x]\, dx$

j) $\displaystyle\int_0^t \sin[4\,x]\,dx$

k) $\displaystyle\int_0^{\pi/2} e^{\sin[x]}\,\cos[x]\,dx$

l) $\displaystyle\int_1^t \left(\frac{1}{x}\right)\,dx$

m) $\displaystyle\int_1^t \left(\frac{1}{x^2}\right)\,dx$

n) $\displaystyle\int_a^t \frac{f'[x]}{f[x]}\,dx$

L.4) Here is the plot of a certain function $g[x]$:

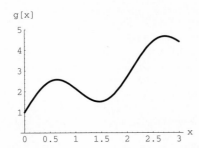

Explain how the plot of $g[x]$ signals that the plot of
$$f[t] = \int_0^t g[x]\,dx$$
goes up as t advances from 0 to 3.

L.5) Here is the plot of a certain function $g[x]$:

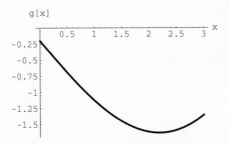

Explain how the plot of $g[x]$ signals that the plot of
$$f[t] = \int_0^t g[x]\,dx$$
goes down as t advances from 0 to 3.

L.6) Here is the plot of a certain function $g[x]$:

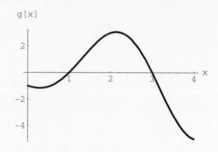

Explain how the plot of $g[x]$ signals that the plot of
$$f[t] = \int_0^t g[x] \, dx$$
$\rightarrow$ goes down as t advances from 0 to 1;

$\rightarrow$ goes up as t advances from 1 to 3; and

$\rightarrow$ goes down as t advances from 3 to 4.

L.7) Here is the plot of a certain function $g[x]$:

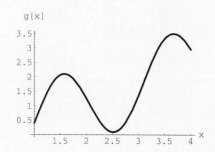

Here are the plots of three other functions:

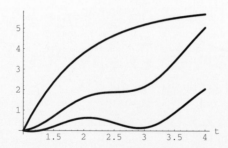

One of the plotted curves is the actual plot of $f[t] = \int_1^t g[x]\, dx$ where $g[x]$ is the function plotted immediately above and the other two are bogus. Pick out the genuine plot of $f[t]$ and say how you came to your decision.

L.8) When you calculate

$$\int_0^t e^{x \cos[2x]}\, dx$$

numerically for various values of t, you get different numbers depending on what t you go with:

In[5]:=
```
Clear[x]; t = 1;
NIntegrate[E^(x Cos[2 x]),{x,0,t},AccuracyGoal->2]
```

Out[5]=
```
1.12294
```

In[6]:=
```
t = 2;
NIntegrate[E^(x Cos[2 x]),{x,0,t},AccuracyGoal->2]
```

Out[6]=
```
1.42524
```

In[7]:=
```
t = 3;
NIntegrate[E^(x Cos[2 x]),{x,0,t},AccuracyGoal->2]
```

Out[7]=
```
5.85576
```

The upshot: When you put

$$f[t] = \int_0^t e^{x \cos[2x]}\, dx,$$

you can plot $f[t]$ as a function of t:

In[8]:=
```
Clear[f,x,t]
f[t_] := NIntegrate[E^(x Cos[2 x]),{x,0,t},
AccuracyGoal->2]; Plot[f[t],{t,0,4},
PlotStyle->Thickness[0.01],
AxesLabel->{"t","f[t]"}];
```

Here's a plot of the solution of the differential equation

$$y'[t] = e^{t \cos[2t]} \qquad \text{with } y[0] = 0:$$

```
In[9]:=
    a = 0; b = 4; starter = 0;
    Clear[solution,t,y,fakey]
    solution = NDSolve[{y'[t] == E^(t Cos[2 t]),
    y[0] == starter},y[t],{t,a,b}];
    fakey[t_] = y[t]/.solution[[1]];
    Plot[fakey[t],{t,a,b},PlotStyle->
    Thickness[0.01],AxesLabel->{"t","y[t]"},
    AxesOrigin->{a,starter}];
```

Explain why it is no accident that the plots are dead ringers for each other.

L.9) There is a big connection between the integral

$$\int_a^t f[x]\,dx$$

and the differential equation

$$y'[t] = f[t] \qquad \text{with } y[a] = 0.$$

What is the connection, and why is it important to set $y[a] = 0$ in the differential equation?

L.10) Use your response to the problem immediately above to explain the similarity of the outputs from:

```
In[10]:=
    Clear[f,x,t]; f[x_] = x E^x;
    Expand[Integrate[f[x],{x,a,t}]]

Out[10]=
         t     t
    1 - E  + E  t

In[11]:=
    Clear[y]
    DSolve[{y'[t] == f[t],y[a] == 0},y[t],t]

Out[11]=
               t    t
    {{y[t] -> 1 - E  + E  t}}
```

L.11) Give a clean formula for $f[x]$ given that

$$f'[x] = \sin[x] \qquad \text{and} \qquad f[0] = \pi.$$

Give a clean formula for $g[x]$ given that

$$g[x] = \pi + \int_0^x \sin[t]\,dt.$$

L.12) Give a clean formula for $f'[x]$ in the case that

$$f[x] = \int_0^x \frac{e^t}{1+2t}\,dt.$$

L.13) Exactly 10 years ago, you owed $1000. At any time t (measured in years with $t = 0$ corresponding to 10 years ago), the amount you owed increased at the rate of

$$500\, e^{t/10} \text{ dollars per year.}$$

About much do you owe now? You can use the estimate of 2.7 for e.

L.14) Calculate

$$\int_0^\infty e^{-x}\, dx$$

L.15) Is this correct:

$$\int_0^2 f[x]\, dx + \int_2^1 f[x]\, dx = \int_0^1 f[x]\, dx?$$

Why or why not?

L.16) Calculus Cal said: "When you go with a constant k and a function $f[x]$, you know that

$$D[f[x], x] = D[f[x] + k, x].$$

As a result, the fundamental formula tells you that when you go with a function $f[x]$ and a constant k, then

$$\int_a^b f[x]\, dx = \int_a^b (f[x] + k)\, dx."$$

What do you say?

L.17) Why would it be a mark of calculus illiteracy to write down

$$\int_{-1}^1 \left(\frac{1}{x^2}\right) dx?$$

L.18) If an object moves on a straight line and is $f[t]$ units away from a reference marker at time t, then its velocity is given by $\text{vel}[t] = f'[t]$ and its acceleration is given by $\text{accel}[t] = \text{vel}'[t]$.

Use the fundamental formula to explain why

$$f[t] = f[a] + \int_a^t \text{vel}[t]\, dt$$

and

$$\text{vel}[t] = \text{vel}[a] + \int_a^t \text{accel}[x]\, dx.$$

L.19) An object moves on a straight line and is $f[t]$ units away from a reference marker at time $t \geq 0$. Its acceleration at time t is given to be

$$\text{accel}[t] = 4 \sin[t] - 32$$

with

$$\text{vel}[0] = 8$$

and

$$f[0] = 12.$$

Come up with formulas for vel[t] and f[t] at any time $t \geq 0$.

L.20) George Will, the nationally syndicated columnist, ABC news analyst, and big-time baseball fan, has a well-deserved reputation as a literate man. In his newspaper column for October 4, 1992, Will wrote: "US Productivity—output per worker—is still higher than Germany's or Japan's, but its growth rate has slowed while productivity has accelerated in other major countries."

$\rightarrow$ Interpret Will's sentence in terms of three functions

$$\text{us}[t], \quad \text{germ}[t], \quad \text{and} \quad \text{japan}[t],$$

their first derivatives

$$\text{us}'[t], \quad \text{germ}'[t], \quad \text{and} \quad \text{japan}'[t]$$

and their second derivatives

$$\text{us}''[t], \quad \text{germ}''[t], \quad \text{and} \quad \text{japan}''[t].$$

$\rightarrow$ Discuss how Will's sentence indicates that in addition to being an accomplished writer, Will knows some calculus.

L.21) Writing

$$\int_a^b x^3 \, dx = \frac{x^4}{4} + C$$

would be a sure sign of calculus illiteracy. Why?

L.22) If you said that

$$\int_a^b (f[x] + g[x]) \, dx = \int_a^b f[x] \, dx + \int_a^b g[x] \, dx,$$

then you would be right. Would you be right if you said that

$$\int_a^b (f[x] \, g[x]) \, dx = \left(\int_a^b f[x] \, dx \right) \left(\int_a^b g[x] \, dx \right)?$$

L.23) Measure the area that is under the curve $y = x$ and is over the curve $y = x^2$.

2.03 Measurements Literacy Sheets

L.1) When you are fresh, you find that you can harvest 150 bushels of corn per hour. But as the day wears on, your efficiency is somewhat decreased. In fact, after t hours from the beginning of the day, you find that you are harvesting at a rate of

$$150\, e^{-0.5t} \text{ bushels per hour.}$$

Write down the integral that measures how many bushels of corn you harvest after arriving in the field fresh as a daisy and working for five consecutive hours.

L.2) At this moment, you borrow \$10,000 at an interest rate of 7% per year compounded every instant. Assuming you borrow no new money and that you pay off none of the loan for the next three years, write down the integral that measures what you owe three years from now.

L.3) You have a profit-making scheme that is projected to pay profits at a rate of

$$p[t] = 100000\, e^{-2t} \text{ dollars per year}$$

t years from now. Assuming a projected interest rate of 6% per year compounded every instant, what is the present value of your scheme?

L.4) The base of a certain solid is the circle

$$x^2 + y^2 = 4$$

in the xy-plane. Each cross section cut by a plane perpendicular to the x-axis is a square. Measure the volume of the solid.

L.5) Quick with no calculation: How long is the circumference of a circle of radius r? What does the area enclosed by a circle of radius r measure out to?

L.6) A certain cone has a planar base whose area measures out to 100 square units, and the height of this cone is 60 units. Measure the volume inside the cone.

L.7) Here's a curve in true scale:

```
In[1]:=
  Clear[x,y,t]
  x[t_] = 1.5 (Sin[t] + Sin[3 t]/3);
  y[t_] = 1 + Cos[t] - Cos[2 t]/2;
  ParametricPlot[{x[t],y[t]},{t,0.2, 2 Pi - 0.2},
  PlotStyle->Thickness[0.01],
  AspectRatio->Automatic,
  PlotRange->All,AxesLabel->{"x","y"}];
```

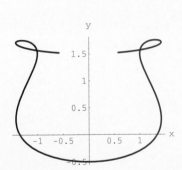

Write down the integral that you would feed into NIntegrate to measure the length of this beauty.

L.8) Here's the curve above with the gap filled:

```
In[2]:=
  ParametricPlot[{x[t],y[t]},
  {t,0, 2 Pi},
  PlotStyle->Thickness[0.01],
  PlotRange->All,
  AspectRatio->Automatic,
  AxesLabel->{"x","y"}];
```

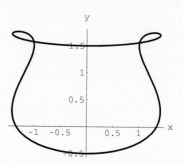

Write down the integral that you would feed into NIntegrate to measure the length of this curve.

L.9) When you're given functions $x[t]$ and $y[t]$ and you're given numbers a and b, you get a curve by plotting out the points $\{x[t], y[t]\}$ as t advances from a to b. In fact, the integral

$$\int_a^b \sqrt{x'[t]^2 + y'[t]^2}\, dt$$

measures the actual length of the plotted curve. If you take any s with $a < s < b$, you get a part of the same curve by plotting out the points $\{x[t], y[t]\}$ as t advances from a to s. In fact, the integral

$$\text{length}[s] = \int_a^s \sqrt{x'[t]^2 + y'[t]^2}\, dt$$

measures the length of this part of the original curve. Write down a clean formula for $\text{length}'[s]$ in terms of $x'[s]$ and $y'[s]$.

L.10) Here's the circle $x^2 + y^2 = 1$ shown with the ellipse

$$\left(\frac{x}{3}\right)^2 + y^2 = 1:$$

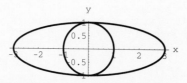

Now look at some horizontal bars:

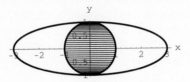 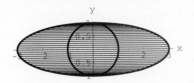

The bars hit the y-axis in the same spots in both plots, but the bars in the second plot are three times longer than in the first plot.

Given that the area inside the circle $x^2 + y^2 = 1$ measures out to π square units, how do the plots reveal the measurement of the area inside the ellipse

$$\left(\frac{x}{3}\right)^2 + y^2 = 1?$$

L.11)   Measure the area inside the ellipse

$$\left(\frac{x}{3}\right)^2 + \left(\frac{y}{2}\right)^2 = 1.$$

Then measure the area inside the ellipse

$$\left(\frac{x}{a}\right)^2 + \left(\frac{y}{b}\right)^2 = 1$$

where a and b are given positive constants.

L.12)   Here is a modest curve shown with the three coordinate axes in three dimensions:

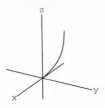

Here is the tube consisting all circles of radius 2 parallel to the xy-plane and centered on the curve:

When you move each circle in its own plane so that it is centered on the z-axis, you get this:

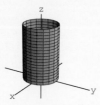

The cylinder has a base with radius 2, and the height of the cylinder is 3; so the volume of the cylinder measures out to

$$2^2 \pi 3 = 12\pi \text{ cubic units.}$$

Explain why the volume of the tube also measures out to 12π cubic units.

L.13) A region is sitting in the xy-plane. All of its linear dimensions are doubled. What is the ratio of the new area to the original area?

L.14) All the linear dimensions of a certain solid are doubled. What is the ratio of the new volume to the original volume? What is the ratio of the new surface area to the original surface area?

L.15) The volume of a sphere of radius 1 is $4\pi/3$ cubic units, and the surface area of a sphere of radius 1 is 4π cubic units. Use the idea of linear dimension to give formulas for the surface area and the volume of a sphere of radius r.

L.16) Here is a rectangular metal plate 40 centimeters long, 20 centimeters wide, and 8 centimeters thick with a cross section y centimeters from the left end and parallel to the left end:

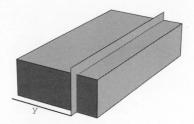

When you examine the cross section y centimeters from the left end, you find that the density is the same throughout the whole cross section and that the density is given by

$$\text{density}[y] = (y + 1)(49 - y) = 49 + 48y - y$$

in grams per cubic centimeter. Measure the mass (in grams) of the plate.

2.04 Transforming Integrals Literacy Sheets

L.1) Explain how to transform the integral

$$\int_0^3 \cos[x^2]\, 2\, x\, dx \qquad \text{into} \qquad \int_0^9 \cos[u]\, du.$$

Use the second integral to calculate the first.

L.2) Explain how to transform the integral

$$\int_0^5 \cos[x^2]\, x\, dx \qquad \text{into} \qquad \left(\frac{1}{2}\right)\int_0^{25} \cos[u]\, du.$$

Use the second integral to calculate the first.

L.3) Explain how to transform the integral

$$\int_0^{\pi/2} e^{\sin[x]}\, \cos[x]\, dx \qquad \text{into} \qquad \int_0^1 e^t\, dt.$$

Use the second integral to calculate the first.

L.4) Explain how to transform the integral

$$\int_0^{\log[3]} \frac{e^x}{1+e^x}\, dx \qquad \text{into} \qquad \int_2^4 \left(\frac{1}{t}\right)\, dt.$$

Use the second integral to calculate the first.

L.5) Use a transformation to help calculate

$$\int_0^{\log[5]} \frac{e^{-x}}{1+e^{-x}}\, dx$$

by hand.

L.6) Here are three integrals:

$$\int_0^1 e^{-[x^3]}\, 3\, x^2\, dx, \quad \int_0^1 e^{-[x^2]}\, 2\, x\, dx, \quad \text{and} \quad \int_0^1 e^{-x}\, dx.$$

Use transformations to explain why all these integrals calculate out to the same value.

L.7) Here is *Mathematica*'s calculation of

$$\int_0^{\log[2]} \sin\left[\left(\frac{\pi}{2}\right) e^{2x}\right] e^x\, dx :$$

In[1]:=
```
Integrate[Sin[(Pi/2) E^(2 x)] E^x,{x,0,Log[2]}]
```

Out[1]=
  -FresnelS[1] + FresnelS[2]

Remembering that

$$\text{FresnelS}[x] = \int_0^x \sin\left[\left(\frac{\pi}{2}\right) t^2\right] dt,$$

use a transformation to explain *Mathematica*'s calculation.

L.8) The basic bell-shaped curve is the plot of

$$\text{bell}[x] = \left(\frac{1}{\sqrt{2\pi}}\right) e^{-x^2/2}.$$

Sketch this curve on the axes below:

The plotted points are on the curve.

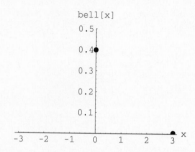

L.9) The function normal[x,mean,dev] is defined by

$$\text{normal}[x, \text{mean}, \text{dev}] = \frac{\text{bell}[(x - \text{mean})/\text{dev}]}{\text{dev}}.$$

Here are plots of normal[x, 50, 10] and normal[x, 70, 10].

Which is the plot of normal[x, 50, 10], and how do you know?

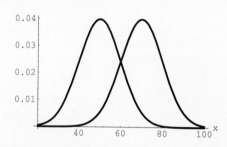

L.10) Here are plots of normal[x, 50, 10] and normal[x, 50, 5]. Which is the plot of normal[x, 50, 5], and how do you know?

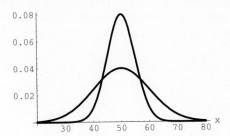

L.11) One group of measurements is normally distributed with mean 80 and standard deviation 10. Another group of measurements is normally distributed with mean 80 and standard deviation 3. Which group of measurements is more tightly packed around 80?

L.12) Use a transformation to explain why

$$\int_{\text{mean}-s\,\text{dev}}^{\text{mean}+s\,\text{dev}} \text{normal}[x, \text{mean}, \text{dev}] = \int_{-s}^{s} \text{bell}[x]\,dx.$$

L.13) Here's a little table of values of $\int_{-s}^{s} \text{bell}[x]\,dx$ for selected values of s:

$$s = 0.5 \longrightarrow \int_{-s}^{s} \text{bell}[x]\,dx = 0.383,$$

$$s = 1.0 \longrightarrow \int_{-s}^{s} \text{bell}[x]\,dx = 0.683,$$

$$s = 1.5 \longrightarrow \int_{-s}^{s} \text{bell}[x]\,dx = 0.866,$$

$$s = 2.0 \longrightarrow \int_{-s}^{s} \text{bell}[x]\,dx = 0.954,$$

$$s = 2.5 \longrightarrow \int_{-s}^{s} \text{bell}[x]\,dx = 0.988.$$

Test scores on a statewide test turn out to be approximately normally distributed with mean 70 and standard deviation 10.

About what percentage of the scores turned out between $70 - 5$ and $70 + 5$?

About what percentage of the scores turned out between $70 - 10$ and $70 + 10$?

About what percentage of the scores turned out between $70 - 15$ and $70 + 15$?

About what percentage of the scores turned out between $70 - 20$ and $70 + 20$?

About what percentage of the scores turned out between $70 - 25$ and $70 + 25$?

About what percentage of the scores turned out over 90?

About what percentage of the scores turned out over 80?

About what percentage of the scores turned out over 70?

About what percentage of the scores turned out below 60?

About what percentage of the scores turned out below 50?

L.14) When you collect data in the form of numerical measurements, how do you recognize whether the data are normally distributed? If you decide that the data are normally distributed, how do you try to estimate mean and dev for the collected data?

L.15) Here's a closed curve in true scale:

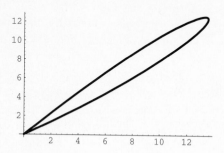

This curve is traced out by the parametric formula

$$\{x[t], y[t]\} = \{6\ t\ (3-t), t\ (3-t)\ (t+4)\} = \{18\,t - 6\,t^2,\ 12\,t - t^2 - t^3\}$$

as t advances from 0 to 3. No part of the curve is traced out more than once. Use an integral to measure the area enclosed by this curve.

Determine whether the curve is traced out in the clockwise or the counterclockwise fashion as t advances from 0 to 3.

L.16) The usual way of specifying a point in the plane is to give its coordinates $\{x, y\}$.

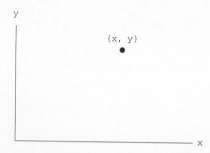

You can specify the same point with the polar angle t and the distance r from the point to the origin. Indicate on this plot what r and t are.

2.05 2D Integrals and the Gauss-Green Formula Literacy Sheets

L.1) Here's a plot of a certain surface

$$z = f[x, y]$$

over the rectangle R in the xy-plane consisting of all points $\{x, y, 0\}$ with $-2 \le x \le 2$ and $-1 \le y \le 3$:

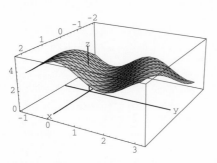

Here's an embellishment of the same plot:

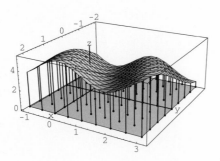

Each of the vertical plotted poles has its base at a point $\{x, y, 0\}$ in the rectangle R. What does the length of each plotted pole measure?

L.2) Given a specific function $f[x, y]$, how do you expect the output from the *Mathematica* instructions:

```
Plot3D[f[x,y], {x,0,4}, {y, -3, 5}]
```

and

```
ParametricPlot3D[{x, y, f[x,y]}, {x, 0, 4}, {y, -3, 5}]
```

to be related?

L.3) Here's a plot of a certain surface $z = f[x, y]$ over the rectangle R in the xy-plane consisting of all points $\{x, y, 0\}$ with $-1 \leq x \leq 2$ and $-1 \leq y \leq 3$ shown with a plot of R:

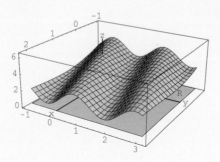

Why does the plot signal that

$$\iint_R f[x, y] \, dx \, dy > 0?$$

L.4) Here's a plot of a certain surface $z = f[x, y]$ over (or under) the rectangle R in the xy-plane consisting of all points $\{x, y, 0\}$ with $-1 \leq x \leq 1$ and $-1 \leq y \leq 2$ shown with a plot of R:

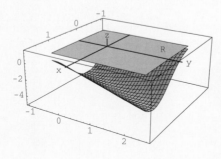

Why does the plot signal that

$$\iint_R f[x, y] \, dx \, dy < 0?$$

L.5) Here's a plot of a certain surface $z = f[x, y]$ over (and under) the rectangle R in the xy-plane consisting of all points $\{x, y, 0\}$ with $-1 \leq x \leq 2$ and $-1 \leq y \leq 3$ shown with a plot of R:

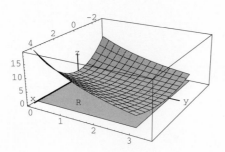

Why does the plot strongly suggest that

$$\iint_R f[x,y]\,dx\,dy > 0?$$

How does the sign of $f[x,y]$ tell you whether the surface plots out over R or under R?

L.6) Here's a plot of the surface

$$z = x^2\,e^{-y/2}$$

over the rectangle R in the xy-plane consisting of all points $\{x,y,0\}$ with $-3 \le x \le 4$ and $0 \le y \le 5$ shown with a plot of R:

Measure by hand calculation the volume of the solid whose top skin is the plotted surface and whose bottom skin is the rectangle R.

L.7) Here's a plot of the surface $z = 3 + \sin[2\,x] + \cos[3\,y]$ over the rectangle R in the xy-plane consisting of all points $\{x,y,0\}$ with $-\pi \le x \le 2\,\pi$ and $-\pi \le y \le 2\,\pi$ shown with a plot of R:

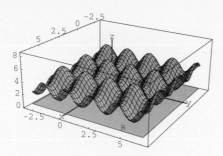

Measure by hand calculation the volume of the solid whose top skin is the plotted surface and whose bottom skin is the rectangle R.

L.8)   Here's a plot of the surface

$$z = 3 - \frac{y^2}{4} - \frac{x}{2}$$

over the region R in the xy-plane consisting of all points $\{x, y, 0\}$ with $\{x, y\}$ inside the 2D circle $x^2 + y^2 = 4$ shown with a plot of R:

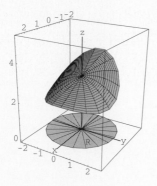

Come up with a counterclockwise parameterization $\{x[t], y[t]\}$ of the circle

$$x^2 + y^2 = 4.$$

Use your parameterization and the Gauss-Green formula to measure by hand calculation the volume of the solid whose top skin is the plotted surface and whose bottom skin is the rectangle R. Don't forget that

$$\int_a^b \sin[t]^k \cos[t] \, dt = \frac{\sin[b]^{k+1}}{k+1} - \frac{\sin[a]^{k+1}}{k+1}$$

and

$$\int_a^b \cos[t]^k \sin[t] \, dt = \frac{-\cos[b]^{k+1}}{k+1} + \frac{\cos[a]^{k+1}}{k+1}.$$

L.9) Go with $f[x] = \sin[2\,x]$. Give a clean formula for the function $h[x]$ defined by

$$h[x] = \int_0^x f[s]\,ds.$$

How are $f[x]$ and $D[h[x], x]$ related? Is the outcome an accident? Why or why not?

L.10) Go with $f[x, y] = 2\,x + 5\,\sin[y]$. Give a clean formula for the function $n[x, y]$ defined by

$$n[x, y] = \int_0^x f[s, y]\,ds.$$

How are $f[x, y]$ and $D[n[x, y], x]$ related?

Is the outcome an accident? Why or why not?

L.11) Go with $f[x, y] = 2\,e^{-x^2}\cos[y]$. Give a clean formula for the function $m[x, y]$ defined by

$$m[x, y] = \int_0^y f[x, s]\,ds.$$

How are $f[x, y]$ and $D[m[x, y], y]$ related?

Is the outcome an accident? Why or why not?

L.12) Write down the Gauss-Green formula and say how to use it to calculate

$$\iint_R x^2\,y^3\,dx\,dy$$

in the case that you have your hands on a convenient counterclockwise 2D parameterization $\{x[t], y[t]\}$, $a \le t \le b$ of the boundary of R.

L.13) Here is a 2D region R with bottom boundary curve

$$\mathrm{low}[x] = x\,(x - 1)$$

and with top boundary curve

$$\mathrm{high}[x] = 2\,x\,(1 - x):$$

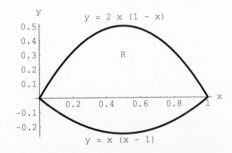

Say why it's natural to calculate

$$\iint_R (x+y)\,dx\,dy$$

by integrating with respect to y first. Next, calculate

$$\iint_R (x+y)\,dx\,dy$$

by hand.

L.14) Here is a 2D region R whose left boundary is the curve

$$\text{left}[y] = \sin[3\,y]$$

and whose right boundary is the curve

$$\text{right}[y] = 4\,\sin[y] :$$

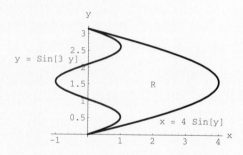

Say why it's natural to calculate

$$\iint_R 1\,dx\,dy$$

by integrating with respect to x first. Next, calculate

$$\iint_R 1\,dx\,dy$$

by hand and say what

$$\iint_R 1\,dx\,dy$$

measures.

L.15) When you want to plot the ellipse

$$\left(\frac{x}{4}\right)^2 + \left(\frac{y}{3}\right)^2 = 1$$

in two dimensions with a counterclockwise parameterization, you can go with:

In[1]:=
```
Clear[x2D,y2D,t]
{x2D[t_],y2D[t_]} = {4 Cos[t],3 Sin[t]};
ParametricPlot[{x2D[t],y2D[t]},{t,0,2 Pi},
PlotStyle->Thickness[0.01],
AspectRatio->Automatic,
AxesLabel->{"x","y"}];
```

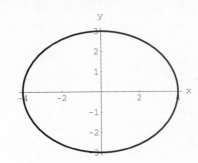

When you want to plot the same ellipse

$$\left(\frac{x}{4}\right)^2 + \left(\frac{y}{3}\right)^2 = 1$$

and everything inside it on the xy-plane in three dimensions with a counterclockwise parameterization, you can go with:

In[2]:=
```
Clear[x,y,z,t]
{x[r_,t_],y[r_,t_],z[r_,t_]} =
r {4 Cos[t],3 Sin[t],0};
Rplot = ParametricPlot3D[
{x[r,t],y[r,t],z[r,t]},{r,0,1},{t,0,2 Pi},
PlotPoints->{2,Automatic},
DisplayFunction->Identity];
spacer = 0.2; threedims = Graphics3D[{
{Line[{{-5,0,0},{5.5,0,0}}]},
Text["x",{5.5 + spacer,0,0}],
{Line[{{0,-3,0},{0,3.5,0}}]},
Text["y",{0, 3.5 + spacer,0}],
{Line[{{0,0,0},{0,0,2}}]},
Text["z",{0,0,2 + spacer}]}];
CMView = {2.7,1.6,1.2}; all = Show[Rplot,threedims,
BoxRatios->Automatic,ViewPoint->CMView,
PlotRange->All,DisplayFunction->$DisplayFunction];
```

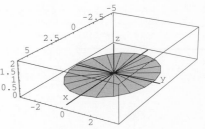

Now look at this:

In[3]:=
```
{{x2D[t],y2D[t]},{x[r,t],y[r,t],z[r,t]}}
```

Out[3]=
```
{{4 Cos[t], 3 Sin[t]}, {4 r Cos[t], 3 r Sin[t], 0}}
```

Where do the points $\{x[r,t],\ y[r,t],\ z[r,t]\}$ land when you go with $r = 0$?

Where do the points $\{x[r,t],\ y[r,t],\ z[r,t]\}$ land when you go with $0 < r < 1$?

Where do the points $\{x[r,t],\ y[r,t],\ z[r,t]\}$ land when you go with $r = 1$?

L.16) Here's the surface

$$z = f[x, y] = e^{(x/5)^2 + (y/3)^2}$$

plotted above the ellipse in L.15):

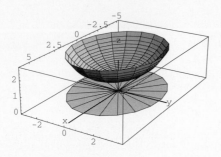

Where do the points $\{x[r, t], \ y[r, t], \ f[x[r, t], \ y[r, t]]\}$ land when you go with $r = 0$?

Where do the points $\{x[r, t], \ y[r, t], \ f[x[r, t], \ y[r, t]]\}$ land when you go with $0 < r < 1$?

Where do the points $\{x[r, t], \ y[r, t], \ f[x[r, t], \ y[r, t]]\}$ land when you go with $r = 1$?

2.06 More Tools and Measurements Literacy Sheets

L.1) Give a formula for the solution $y[x]$ of the differential equation

$$y'[x] = 0.4 \, y[x] \qquad \text{with } y[0] = 8.$$

L.2) Give a formula for the solution $y[x]$ of the differential equation

$$y'[x] = \frac{x^3}{y[x]} \qquad \text{with } y[0] = 1.$$

L.3) Why won't the method of separation of variables and integrating work to obtain a formula for $y[x]$ given

$$y'[x] = x^2 \, y[x] + 4 \, \sin[x] \qquad \text{with } y[0] = 3?$$

L.4) Use separation of variables to derive the formula $y[x] = k \, e^{rx}$ for a function $y[x]$ solving

$$y'[x] = r \, y[x] \qquad \text{with } y[0] = k.$$

L.5) Comment on the quotation:

"The derivative does not display its full strength until it is allied with the integral."

... Michael Spivak

L.6) Calculate the following by integration by parts:

a) $\displaystyle \int_1^e \log[x] \, dx$

b) $\displaystyle \int_1^\infty x \, e^{-2x} \, dx$

c) $\displaystyle \int_1^5 \text{SinIntegral}[x] \, dx \qquad \left(\text{SinIntegral}[x] = \int_o^x \left(\frac{\sin[t]}{t} \right) \, dt \right)$

L.7) What is the integration by parts formula and how is it related to the product rule of differentiation?

L.8) How is the complex number i related to -1?

L.9) What is the value of of $e^{i\pi}$?

How does the value of $e^{i\pi}$ force $\log[-1] = i \, \pi$?

Give the value of $\log[-4] = \log[4] + \log[-1]$.

L.10) Where do these formulas come from?

$$e^{imt} = \cos[m \, t] + i \, \sin[m \, t]$$
$$e^{imt} = (\cos[t] + i \, \sin[t])^m$$
$$e^{-imt} = \cos[m \, t] - i \, \sin[m \, t]$$
$$e^{-imt} = (\cos[t] - i \, \sin[t])^m$$

$$\cos[m\,t] = \frac{e^{imt} + e^{-imt}}{2}$$

$$\cos[m\,t] = \frac{(\cos[t] + i\,\sin[t])^m + (\cos[t] - i\,\sin[t]^m)}{2}$$

How do you use the last formula to slam out formulas for $\cos[m\,t]$ in terms of $\sin[t]$ and $\cos[t]$?

What did the great scientist Richard Feynman mean when he said, "The algebra of [the complex] exponentials is much easier than that of sines and cosines"?

L.11) Write some words about the hyperbolic functions $\sinh[x]$ and $\cosh[x]$ and their relation to the trig functions $\sin[x]$ and $\cos[x]$.

L.12) Set

$$\text{Int}[n] = \int_0^\infty x^n\,e^{-x}dx$$

and find a formula that expresses $\text{Int}[n]$ in terms of $\text{Int}[n-1]$.

Use your formula and a pencil to write out the values of $\text{Int}[0]$, $\text{Int}[1]$, $\text{Int}[2]$, $\text{Int}[3]$, and $\text{Int}[4]$.

2.07 Traditional Pat Integration Procedures for Special Situations Literacy Sheets

This lesson is not part of the mainline course.

L.1) Calculate

$$\int_0^t \cos[x]^4 \, \sin[x] \, dx.$$

L.2) Use the identity

$$\sin[x]^2 = \frac{1 - \cos[2\,x]}{2}$$

to help calculate

$$\int_0^t \sin[x]^2 \, dx.$$

L.3) Calculate

$$\int_0^t x \, e^{-x} \, dx.$$

L.4) Calculate

$$\int_2^3 \frac{1}{(x-1)\,(x-4)} \, dx.$$

L.5) Calculate

$$\int_0^3 \frac{1}{(9 - x^2)^{1/2}} \, dx.$$

L.6) Calculate

$$\int_0^t \frac{e^{-x}}{1 + e^{-x}} \, dx.$$

Index

Entries are listed by lesson number and problem number.

Acceleration, measuring via fundamental
    formula, 2.02: T.2, G.3
Accumulated growth measurements, 2.03:
    B.4, G.1
Accumulation, measurements based on. *See*
    Measurements
Approximation
    integration by, by trapezoids,
        2.01: B.3
    measurements based on, arc length,
        2.03: B.3
Arc length
    derivative of, 2.03: G.7
    measurement of, 2.03: B.3
Area measurements, 2.03: B.1, T.2
    of area between curves, measuring via
        fundamental formula, 2.02: T.4
    integrals for. *See also* Integrals, for
        measuring area
        transforming, 2.04: T.4, G.3, G.7
        2D, 2.05: T.1, G.3
    polar plots and, 2.04: T.4
    slicing for, 2.03: G.3
Average value, 2D integrals and, 2.05: G.6

Backward integration, 2.02: B.6
Bell-shaped curve, 2.04: B.3
Bombing runs, planning with 2D integrals,
    2.05: T.5, G.7

Catfish harvesting, 2.03: G.10
Centroids, 2D integrals and, 2.05: G.6

Champagne glasses, design of,
    2.03: G.6
Chemical reaction model, 2.06: G.3
Circles, clockwise versus counterclockwise
    parametrization and, transforming
    integrals and, 2.04: G.11
Coefficients, undetermined, 2.07: B.2
Complex exponentials
    algebra of, 2.06: G.7
    Mathematica output from Solve instruction
        and, 2.06: T.3
Complex numbers, 2.06: B.3, 2.07: T.3
cosh[x], 2.06: G.5
Cosines, algebra of, 2.06: G.7
Curves
    area between, measuring via fundamental
        formula, 2.02: T.4
    area under, given parametrically, measuring,
        2.04: B.2
    closed, area inside, measurement of,
        2.04: T.3

Data lists, areas suggested by, integration of,
    2.01: T.2, G.8
Density measurements, 2.03: B.2
Derivative, of arc length, 2.03: G.7
Differential equations
    calculating integrals by solving, 2.02: G.1
    fundamental formula and, 2.02: T.6
    separating and integrating variables to
        solve, 2.06: B.1, T.1, G.1

Distance measurements, fundamental formula and, 2.02: B.3

$e^{a+ib} = e^a(\cos[b] = i\sin[b])$, 2.06: B.3

Equations, differential. *See* Differential equations
Error propagation, iteration and, 2.06: G.8
Exponentials, complex
  algebra of, 2.06: G.7
  Mathematica output from Solve instruction and, 2.06: T.3

Fractions, partial, 2.07: T.3
  for quotients of polynomials, 2.07: B.3
Functions, defined by integrals, 2.02: G.4
Fundamental formula
  approximate calculation of $\int_a^\infty f[x]dx$ and, 2.02: T.5
  area between curves and, 2.02: T.4
  backward integration and, 2.02: B.6
  calculating $f'[t]$ when $f[t] = \int_a^t g[x]dx$, 2.02: B.1
  calculating integrals using, 2.02: T.1
  Mathematica and hand calculations and, 2.02: G.2
  by solving differential equations, 2.02: G.1
  differential equations related to, 2.02: T.6
  distance and velocity measurements via, 2.02: B.3
  exact and approximate calculations of $\int_a^\infty f[x]dx$ and, 2.02: G.7
  failure of *Mathematica* to calculate, 2.02: G.9
  failures of, 2.02: G.9
  $f[t] - f[a] = \int_a^t f'[x]dx$, 2.02: B.2
  functions defined by integrals and, 2.02: G.4
  $\int_a^\infty f[x]dx$ and, 2.02: B.4
  indefinite integral and, 2.02: T.7
  integral of sum as sum of integrals and, 2.02: B.5
  measurements based on, 2.02: T.3, G.6
  of accumulated growth, 2.03: B.4
  plotting $(f[t+h] - f[t])/h$ and $g[t]$ when $f[t] = \int_a^t g[x]dx$ and, 2.02: G.5
  velocity and acceleration and, 2.02: T.2, G.3
  Waterloo tiles and, 2.02: G.8

Gamma function, 2.06: G.6
Gauss-Green formula
  area and volume measurements and, 2.05: G.3
  calculating $\int_R f[x,y]\,dx\,dy$ and, 2.05: B.4

  clockwise parameterization and, 2.05: T.4
  ideas behind, 2.05: B.5
Gauss's normal law
  transforming integrals and, 2.04: B.3, T.5, G.5, G.9
  in 2D, 2.05: T.5
  planning bombing runs and, 2.05: T.5
Growth, accumulated, measurement of, 2.03: B.4, G.1
Guessing
  partial fractions for quotients of polynomials, 2.07: B.3
  undetermined coefficients and, 2.07: B.2

Indefinite integral, 2.02: T.7
Integrals
  functions defined by, 2.02: G.4
  fundamental formula of. *See* Fundamental formula
  indefinite, 2.02: T.7
  for measuring area, 2.01: B.1
    calculating, 2.01: T.1, G.1, G.2, G.3, G.9
    estimating and, 2.01: G.7
    experiments to break code of integral, 2.01: G.4
    guessing and, 2.01: G.9
    integration by approximation by trapezoids and NIntegrate instruction and, 2.01: B.3
    integration of data lists and, 2.01: T.2, G.8
    measuring and, 2.01: G.9
    NIntegrate instruction and, 2.01: G.6; B.3
    nonsense integrals, 2.01: T.3
    numerical integration, 2.01: G.5
    plotting, 2.01: G.1, G.2, G.9
    properties of, 2.01: B.2
    symmetry and, 2.01: G.1
    transforming integrals and. *See* Integrals, transforming
  properties of, 2.01: B.2
  of sum, as sum of integrals, 2.02: B.5
  tables of, iteration and, 2.06: G.4
  transforming, 2.04: B.1, T.1, G.1
    bell-shaped curves and, 2.04: B.3, G.6, G.9
    clockwise versus counterclockwise parameterization and, 2.04: G.11
    explaining measurements of area, length and volume and, 2.04: G.7
    Gauss's normal law and, 2.04: B.3, T.5, G.5

Integrals (*continued*)
in materials science, 2.04: G.8
measuring area under curves parametrically
and, 2.04: B.2, T.3, G.3
polar plots and area measurements and,
2.04: T.4, G.10
understanding *Mathematica* output and,
2.04: T.2, G.2
volume measurements and, 2.04: G.4
work and velocity and, 2.04: G.12
trigonometric, 2.07: B.4
2D. *See* 2D integrals
Integration. *See also* Fundamental formula;
Integrals; Pat integration procedures;
2D integrals
backward, 2.02: B.6
by iteration, 2.06: T.2
by parts, 2.06: B.2, T.2, G.2, 2.07: B.6
to solve differential equations, 2.06: B.1,
T.1, G.1
Iteration
error propagation via, 2.06: G.8
integration by, 2.06: T.2
tables of integrals by, 2.06: G.4

Length measurements, 2.03: G.2
transforming integrals and, 2.04: G.7
Linear dimensions, measurement of, 2.03:
T.2, G.9
Logarithm(s), of negative numbers, 2.06: B.3
Logistic differential equation, solution of,
2.06: B.1

Mass measurements, 2.03: B.2, G.2
Materials science, transforming integrals and,
2.04: G.8
Measurements. *See also specific types of
measurements, i.e.,* Area measurements
of accumulation, 2.03: G.1
based on approximating and accumulating,
arc length, 2.03: B.3
based on fundamental formula, 2.02: T.3,
G.6
accumulated growth, 2.03: B.4
based on slicing and accumulating
area and volume, 2.03: B.1
density and mass, 2.03: B.2
catfish harvesting and, 2.03: G.10
derivative or arc length and, 2.03: G.7
of distance, fundamental formula and,
2.02: B.3
of length, volume, and mass, 2.03: G.2
of linear dimensions, 2.03: G.9
volume and area, 2.03: T.2

normally distributed, transforming
integrals and, 2.04: B.3, G.6, G.9
present value of a profit-making scheme
and, 2.03: G.8
slicing for area measurements and,
2.03: G.3
of velocity, fundamental formula and,
2.02: B.3
volume measurements for tubes and horns,
2.03: G.4
of volumes of solids, 2.03: T.1
of work, 2.03: G.5, 2.04: G.12

Negative numbers, logarithms of, 2.06: B.3
NIntegrate instruction, 2.01: B.3, G.6
Nonsense integrals, 2.01: T.3
Normal distribution, transforming integrals
and, 2.04: B.3, G.6, G.9
Normally distributed measurements, 2.04:
B.3, T.5, G.5, G.9

Parametric 3D plots, 3D plots versus,
2.05: G.4
Partial fractions, 2.07: T.3
for quotients of polynomials, 2.07: B.3
Pat integration procedures
complex numbers and, 2.07: T.3
definition and role of, 2.07: B.1
integrals to try and, 2.07: T.2
integration by parts, 2.07: B.6
partial fractions and, 2.07: T.3
for quotients of polynomials, 2.07: B.3
traditional, 2.07: T.1
trigonometric integrals, 2.07: B.4
undetermined coefficients, 2.07: B.2
wild card substitution, 2.07: B.5
Plotting, of integrals. *See* Integrals
Polar plots
area measurements and, 2.04: T.4
rotating and measuring, transforming
integrals and, 2.04: G.10
Polynomials, quotients of, partial fractions
for, 2.07: B.3
Present value, 2.03: G.8

Quotients, of polynomials, partial fractions
for, 2.07: B.3

Sines, algebra of, 2.06: G.7
sinh[x], 2.06: G.5
Slicing, measurements based on. *See*
Measurements
Solve instruction, complex exponential and,
2.06: T.3

Spread of infection model, 2.06: G.3

Symmetry, integrals and, 2.01: G.1

3D plots, parametric 3D plots versus,
2.05: G.4

Transformation, of integrals. *See* Integrals,
transforming

Trapezoids, integration by approximation by,
2.01: B.3

Trigonometric integrals, 2.07: B.4

2D integrals
area measurements using, 2.05: T.1
Gauss-Green formula and, 2.05: G.3
average value and centroids and, 2.05: G.6
calculating, 2.05: G.2
calculating $\iint_R f[x, y]\,dx\,dy$ and
Gauss-Green formula and, 2.05: B.4, B.5
when R isn't a rectangle, 2.05: B.3
calculation strategies for, 2.05: T.3
Gauss-Green formula and, 2.05: B.4
with clockwise parameterization,
2.05: T.4
ideas behind, 2.05: B.5
volume measurement and, 2.05: G.3
Gauss's normal law and, planning bombing
runs using, 2.05: T.5
planning bombing runs using, 2.05: G.7
Gauss's normal law and, 2.05: T.5

plotting and measuring with, 2.05: G.5
$\iint_R f[x, y]\,dx\,dy$ when $f[x, y]$ isn't always
positive and, 2.05: B.2
3D plots versus parametric 3D plots and,
2.05: G.4
volume measurements using, 2.05: B.1,
T.2, G.1
Gauss-Green formula and, 2.05: G.3

Undetermined coefficients, 2.07: B.2

Variables, separating and integrating to solve
differential equations, 2.06: B.1, T.1,
G.1
Velocity measurements
fundamental formula and, 2.02: B.3, T.2,
G.3
transforming integrals and, 2.04: G.12
Volume measurements, 2.03: B.1, T.1, T.2,
G.2, G.4
transforming integrals and, 2.04: G.4, G.7
2D integrals for, 2.05: B.1, T.2, G.1
Gauss-Green formula and, 2.05: G.3

Waterloo tiles, 2.02: G.8

Wild card substitution, 2.07: B.5

Work, measurement of, 2.03: G.5
transforming integrals and, 2.04: G.12